Microcomputer-Based Adaptive Control Applied to Thyristor-Driven DC-Motors

Other titles published in this Series:

Parallel Processing for Jet Engine Control
Haydn A. Thompson

Iterative Learning Control for Deterministic Systems
Kevin L. Moore

Parallel Processing in Digital Control
D. Fabian Garcia Nocetti and Peter J. Fleming

Intelligent Seam Tracking for Robotic Welding
Nitin Nayak and Asok Ray

*Identification of Multivariable Industrial Processes
for Simulation, Diagnosis and Control*
Yacai Zhu and Ton Backx

Nonlinear Process Control: Applications of Generic Model Control
Edited by Peter L. Lee

Ulrich Keuchel and Richard M. Stephan

Microcomputer-Based Adaptive Control Applied to Thyristor- Driven DC-Motors

With 78 Figures

Springer-Verlag
London Berlin Heidelberg New York
Paris Tokyo Hong Kong
Barcelona Budapest

Ulrich Keuchel
Faculty of Electrical Engineering
Ruhr-University Bochum
P.O. Box 102148
D-44780 Bochum, Germany

Richard M. Stephan
UFJR-Universidade Federal do Rio de Janeiro
Department of Electrical Engineering
P.O. Box 68504
Rio de Janeiro
21945-970 Brazil

ISBN-13: 978-1-4471-2078-0   e-ISBN-13: 978-1-4471-2076-6
DOI: 10.1007/978-1-4471-2076-6

British Library Cataloguing in Publication Data
A catalogue record for this book is available from the British Library

Library of Congress Cataloging-in-Publication Data
A catalog record for this book is available from the Library of Congress

69/3830—543210 Printed on acid-free paper

To
Britta, Bruno, Camila, Evelyn, Katrin, Maik,
Marilia and Vanessa

SERIES EDITORS' FOREWORD

The series *Advances in Industrial Control* aims to report and encourage technology transfer in control engineering. The rapid development of control technology impacts all areas of the control discipline. New theory, new controllers, actuators, sensors, new industrial processes, computing methods, applications, philosophies, . . ., new challenges. Much of this development work resides in industrial reports, feasibility study papers and the reports of advanced collaborative projects. The series offers an opportunity for researchers to present an extended exposition of such new work in all aspects of industrial control for wider and rapid dissemination.

The autotune method of Åström and Hägglund had a major impact on the hardware and structure of PID process controllers. However, despite a substantial body of theoretical analysis, progress in transferring the benefits of more general self-tuning methods to industrial devices and processes has been much slower. This volume by Dr's Stephan and Keuchel shows that this type of technology transfer can be achieved and that the more advanced adaptive controllers do give performance benefits over conventional industrial (three term) controllers. The volume also shows the requirements in hardware, the need for software skills and the engineering techniques required to achieve satisfactory results. We hope that by recording their engineering know-how more researchers and industrialists will be encouraged to tap the benefits of advanced self-tuning and adaptive control methods.

July, 1993   Michael J. Grimble and M. A. Johnson,
Industrial Control Centre,
Glasgow, Scotland, U.K.

PREFACE

The origin of the design and application of adaptive control systems lies more than fifty years ago. Today well-established design procedures for guaranteed stability, robustness and parameter convergence are available for adaptive control systems. Because of this advanced theoretical standard one could expect that the practical application of adaptive control systems would already be widespread in industry. Considering the different application areas, however, one has to confess that the industrial acceptance of adaptive control is still very limited. There are several reasons for this situation, which will not be discussed here. It should, however, be mentioned that most theoretical approaches for adaptive control systems are either too far from the practical demands of the user, or are too complex and provide no physical insight into the operation, so that only specialists can understand and apply them. To overcome this situation the design engineer must provide an approach which copes with the realities and is easily understandable. Furthermore, he has to demonstrate the applicability of his approach by experimental applications, which are easy to understand for the user.

This book is intended as a contribution to the technical realization and critical judgement of different adaptive control concepts in order to bridge the gap between practice and theory. As a special example for application the authors have chosen the adaptive speed control of a separately excited DC-motor fed by a single-phase controlled dual converter. This system is characterized by large variations of its parameters and structure. For example, during normal operation, considerable changes occur in the moment of inertia, field excitation and load torque. Furthermore, the dynamic structure of this machine varies. When the current is continuous the motor armature can be regarded as a first order system, but when the current is discontinuous, it changes into a nearly nonlinear gain system. All these variations make the application of adaptive control systems for speed control attractive.

The authors, both highly qualified specialists in their field, give, after a short introductory chapter, an extensive description of the simplified model of the DC-motor, the investigated experimental laboratory system and the developed hardware and software tools (chapters 2 to 4). Chapter 5 provides the basic theory for the implemented adaptive control schemes. The adaptive controller synthesis and realization, described in chapter 6, is finally implemented on a 16-bit single board microcomputer. The experimental results presented in chapter 7 show that the improved cascaded adaptive speed control systems, based on a discrete model reference adaptive control strategy, provide excellent dynamic behaviour for thyristor-driven DC-motors. In particular, stable operation can be guaranteed, even for non-minimum phase-systems.

It seems to me that the authors have succeeded very well in presenting their material in a concise and understandable form, so that not only specialists in the area of adaptive control, but also practising engineers will be able to apply these control schemes in industry. I find this contribution to be one of the very few examples currently available of a systematic application of adaptive control schemes to industrial plants and I hope that it will be widely accepted.

August, 1993   Heinz Unbehauen

Acknowledgements

The authors would like to express their gratitude especially to Professor Unbehauen, head of the Automatic Control Laboratory at Ruhr-University Bochum, for his support and critical attendance during all the projects covered by the contents of this book. Thanks also to Norbert Fabritz, José Katzenstein, Dr Amit Patra, Dr Christian Schmid, Jörg Vössing and all the others, who contributed to the success of our joint work.

August, 1993   Ulrich Keuchel and Richard M. Stephan

CONTENTS

1 Preliminaries 1

1.1 Introduction 1
1.2 Past Work 1
1.3 Outline of Chapters 6
1.4 References 7

**2 A Simple Model for a Thyristor Driven DC-Motor
Considering Continuous and Discontinuous Current Modes** . . . 13

2.1 Introduction 13
2.2 The DC-Motor Model and the Basic Equations 15
2.3 The Model of a Thyristor Driven DC-Motor 17
    2.3.1 Armature Model 17
    2.3.2 Complete Mean-Value Model 20
2.4 Experimental Results 21
2.5 Conclusion 24
2.6 References 25

3 The Experimental Laboratory System 26

3.1 Introduction 26
3.2 Construction 26
    3.2.1 The Fundamental Unit 28
    3.2.2 The Extension Unit 28
3.3 The System Parameters 29
3.4 References 32

4 The Microcomputers: Hardware and Software 33

4.1 Introduction 33
4.2 The Development System and the Single Board Computer . 34
    4.2.1 The Basic Concept 34
    4.2.2 The Single Board Computer (SBC) 35
    4.2.3 The Teachbox 39
4.3 The Monitor Program 41
    4.3.1 The Single Board Computer ROM-BIOS 41
    4.3.2 The Real-Time Scheduler 45
4.4 The Interface to the CADACS System 47
    4.4.1 Overview to the CADACS System 47

4.4.2 The Data Base and Types of System Representation . 48
4.4.3 Tools for Kernel Functions 49
4.4.4 Tools for Analysis and Synthesis 51
4.4.5 Programs for Real-Time Operation 52
4.4.6 The User Interface 54
4.5 The Support Software 55
4.5.1 Generation of Programs with a PC 55
4.5.2 The Real-Time Toolbox 57
4.6 Conclusion 64
4.7 References 65

5 **Basic Theory of the Implemented Adaptive Control** 67

5.1 Introduction 67
5.2 The Model Reference Adaptive Controller 69
5.2.1 Basic Considerations 69
5.2.2 The Control Law 71
5.2.3 Adaptation of Parameters 73
5.3 The Improved Model Reference Adaptive Controller 74
5.3.1 Problems in Standard Model Reference Adaptive
       Control 74
5.3.2 The Linear Control Law 75
5.3.3 Design of the Correction Network 79
5.3.4 Calculation of the Filtered Model Error 80
5.3.5 Estimation Procedures for Plant and Controller
       Parameters 83
5.4 The Adaptive Pole Placement Controller 85
5.4.1 State Feedback Controller and State Observer 85
5.4.2 The Linear Control Law 89
5.4.3 The Adaptive Control Law 93
5.4.4 Implementation Issues 98
5.5 Conclusion 113
5.6 References 114

6 **Controllers Synthesis and Realization** 118

6.1 Introduction 118
6.2 The Commercially Available Analog Controller 118
6.2.1 The Analog Dual-Mode Adaptive Current Controller 120
6.2.2 The Analog PI Speed Controller 121
6.3 The Completely Adaptive Digital Scheme 122
6.3.1 The Digital Dual-Mode Adaptive Current Controller 122
6.3.2 The Model Reference Adaptive Speed Controller . . 125
6.4 The Completely Adaptive Hybrid Scheme 126
6.5 Conclusion 129
6.6 References 129

7 Experimental Results and Comparisons 131

7.1 Introduction 131
7.2 Experimental Modelling 131
7.3 Comparison of the Control Structures 134
7.4 The Inner Current Loop 136
7.5 The Cascade Speed Control Schemes 139
7.6 Conclusion 148
7.7 References 149

8 Survey and Conclusion 150

Subject Index 152

CHAPTER 1
PRELIMINARIES

1.1 Introduction

The advent of silicon controlled rectifiers (SCR) in 1957 enlarged the scope of the technology for speed control of DC-motors. Developments in SCR units for power control have made modern speed control units cheaper and more compact than conventional ones. The new systems have better dynamic response than the older ones. Developments in digital computer technology in the last decades have further increased the field of DC-motor speed control. Microcomputers are nowadays becoming less expensive and so are widely used as dedicated systems for individual control units. Advanced control strategies, as adaptive control, can now be digitalized and implemented with microcomputers. This book will link these new technologies.

1.2 Past Work

The trends in *digital speed control* of DC motors may be traced in the work of McLaren et al. (1975), Schnieder (1977), Claussen and Fromme (1978), Leonhard (1980), Best and Mutschler (1982), Slattery and Wade (1983), Harashima (1983), Bose (1988) etc.

Among the various methods available today (e.g. Bühler, 1978; Weihrich, 1978; Riedo, 1982; Shirakura et al., 1983; Hsu et al., 1984; De et al., 1985; Utkin, 1993) *cascaded speed control schemes* with an inner current control loop and an outer speed control loop represent the classical structure for DC-motor speed control (e.g. Nandam and Sen, 1987). The application of direct speed control, i.e. without the inner current control loop, is limited to servo motors only (Grotstollen, 1977). Joos

Table 1.1 Some proposals for current control of thyristor fed DC-motors

Authors	Year	Ana-log	Digi-tal	Method
Buxbaum	1969	×		Scheduled adaptation based on current feedback.
Golde & Riebschläger	1971	×		Feed forward control to compensate the nonlinear effect in discontinuous conduction.
Ernst & Ströle	1973	×		Scheduled adaptation based on current feedback.
Oumar et al.	1977		×	Scheduled adaptation based on reference current and speed signals.
Bühler	1978		×	Scheduled adaptation based on firing angle and current mean value.
Chan, Chmiel & Plant	1980		×	Scheduled adaptation based on current mean value, firing angle and back voltage feedback.
Ohmae et al.	1980		×	Scheduled adaptation based on current feedback.
Favre	1982		×	Scheduled adaptation based on current feedback.
Magyar et al.	1982		×	Scheduled adaptation based on current feedback.
Fromme	1982		×	Self optimizing digital controller in time domain.
Joos & Goodman	1983	×		Scheduled adaptation based on current feedback.
Holtz & Schwellenberg	1983	×		Feed forward control to compensate the nonlinear effect in discontinuous current mode.
Kennel&Schröder	1983		×	Predictive control.
Grötzbach	1981		×	Predictive control.
Hasegawa et el.	1984		×	Scheduled adaptation based on current feedback.
Ohmae et al.	1986		×	Additional current rate loop.
Tadakuma et al.	1987		×	Deadbeat control.
Hill&Lua	1987	×	×	Stability analysis.
Collings&Wilson	1991		×	Predictive control.

and Barton (1975), Sen (1990), Joos et al. (1992) reviewed the early works on this control problem.

Certain practical difficulties may arise with the classical control structure. One of them is due to the discontinuous nature of the armature current. As long as the current is continuous the armature may be modelled as a first order system, but as the current becomes discontinuous, it exhibits nonlinear gain behaviour (Kümmel, 1965; Buxbaum, 1969; Pelly, 1971).

Furthermore, variations in load torque, moment of inertia of the load and field excitation may also occur. Under these conditions the desirable performance of the system may be either maintained (i) by employing adaptive control techniques or (ii) by incorporating robust control designs. Usually a fast inner current control loop compensates for the armature behaviour and a relatively slow outer control loop compensates for the rest.

The continuous/discontinuous armature current operating modes can be easily detected with inexpensive circuits, it is a common practice to apply the so called "scheduled adaptation" or "parameter scheduling control" principle for the inner current control loop (Parks et al., 1980; Unbehauen, 1985). The scheduled adaptation, that will be also termed here as "dual-mode adaptation", responds as soon as a plant variation is detected by changing the controller parameters correspondingly. This method was introduced by Buxbaum (1969) to control the armature current of thyristor fed DC-motors. Many contibutions and papers, presented in Table 1.1, discuss comprehensively the problem of the current control loop. Among the investigations described here, the dual-mode adaptive current controller concept will be regarded as a standard technique.

When the variations in field excitation, moment of inertia, load torque, etc. can be measured or observed, it is also possible to implement scheduled adaptation for the outer speed control loop, as described by Ströle (1967a,b). Another way of adaptation is possible with an adaptive observer, as proposed by Weihrich and Wohld (1980). Their adaptive observer reconstructs the ratios of excitation flux to inertia and that of torque to inertia, and with these values it is possible to design a controller with scheduled adaptation. The self-tuning approach was studied by Depping and Voits (1982, 1983), Brickwedde (1983) and Balasubramanian and Wong (1986). Fromme (1982) proposed a self-optimizing digital controller in time domain. The model reference adaptive control (MRAC) of the speed of DC-drives was introduced by Raatz (1970) who applied a control scheme based on the MIT-rule (Unbehauen, 1985). Courtiol and Landau (1975) proposed an analog adaptive system based on the hyperstability theory. They used the adaptive model following control (AMFC) approach, where the conditions of perfect model following

3

Table 1.2 Some approaches for adaptive speed control of DC-motors

Authors	Year	Ana-log	Digi-tal	Adaptation method
Model reference adaptive control (MRAC)				
Raatz	1970	×		MIT Rule.
Sinner	1973	×	×	Hyperstability. Adaptive state variable control. Hybrid system.
Courtiol&Landau	1975	×		Hyperstability. Adaptive model following control.
Balestrino et al.	1983	×		Hyperstability. Adaptive model following control.
Lozano&Noriega	1983		×	Hyperstability. No current loop.
Platzer&Kaufman	1984		×	Simulation study based on the adaptation method of Sobel et al. (1982).
Hong&Zohdy	1985			Lyapunov. No current loop. Simulation study.
Naitoh&Tadakuma	1987		×	Linear model following control. Hybrid system.
Tamura	1991		×	MRAC with no overshoot. No current loop.
Self-tuning control				
Depping&Voits	1983, 1982		×	Deadbeat response with adapted set point variation.
Brickwedde	1983		×	Adaptive control applied to the speed control loop. The inner control loop is a commercially available unit.
Balasubramaniam& Wong	1986		×	IP adaptive speed controller.
Other methods				
Weihrich&Wohld	1980	×		Adaptive observer. Stability proved using Lyapunov's method.
Fromme	1982		×	Self optimizing digital controller in time domain.
Kelly	1987		×	Adaptive feed-forward control.
Ohishi et al.	1988	×	×	Observer theory and passive adaptive control.

4

(Erzberger, 1968) must be satisfied. Sinner (1973) used an adaptive state feedback controller described by Landau (1979). In addition to them digital MRAC suggestions were made. Lozano and Noriega (1983) applied an adaptive algorithm with forgetting factor, introduced originally by Lozano and Landau (1981), for the digital speed control of a DC servo motor without inner current control loop. The proposed adaptive controller is not applicable to nonminimum phase plants.

Balestrino et al. (1983) applied a variant of the adaptive model following control of Courtiol and Landau (1975). Platzer and Kaufman (1984) made simulation studies applying the MRAC algorithm proposed by Sobel et al. (1982). Simulation studies were also carried out for a servo motor by Hong and Zohdy (1985). Naitoh and Tadakuma (1987) proposed a combination of the adaptive controller with the linear model following control (LMFC). A linear state feedback plus adaptive feed forward control scheme for DC servo motors was introduced by Kelly (1987). Ohishi et al. (1988) designed a control system based on observer theory and on a passive adaptive control theory. Tamura et al. (1991) applied a modified MRAC scheme to obtain overshoot-free behaviour for the position control of a DC servo motor. Table 1.2 lists the mentioned papers.

In the present book the speed control is based on a discrete-time model reference adaptive control strategy introduced by Hahn (1983). This strategy guarantees stable operation even when the plant is nonminimum phase (Hahn and Unbehauen, 1982), or when the current reference limits are reached (Hahn, 1985). The method was already used in a multivariable version for control of a distillation column (Wiemer et al., 1983) and of a turbo-generator laboratory system (Hahn et al., 1983). Both implementations were done in mini-computers with standard real-time operating systems. The processes were also sufficiently slow as concerns their dynamics, so that the execution time of the algorithm was not a critical factor as in the present experimental application.

In sum the present book is concerned with the development, description and discussion of improved cascade speed control systems for thyristor driven DC-motors. The main features are the following:

- Development of a digital dual-mode parameter scheduling adaptive controller for the inner armature current control loop and of a controller with perpetual adaptation for the outer speed control loop, either based on model reference principle or on linear quadratic optimal pole placement design, thus making the entire control system adaptive.

- Implementation of the above control strategies in a standard state-of-the-art 16-bit single board microcomputer system with arithmetic coprocessor. This implementation requests a development system able to program, test and debug the various control algorithms and a minimal real-time operating system for the on line operation of the adaptive controllers.

- Besides implementing and combining the different control strategies for the control of a DC-motor, the aim was to compare the results of the improved cascade schemes with a commercially available conventional controller.

1.3 Outline of Chapters

The book is organized in the following way: Chapter 2 presents a mean-value model for thyristor fed DC-motors. The relation between the proposed model with the heuristic mean-value model in current use is explained. Moreover, the role of the AC-DC converter in the model of the DC-motor armature is emphasized. The comparison of experimental and simulation results validates the proposed model.

In chapter 3 an experimental laboratory system and its facilities are described. The need for an interface circuit, that makes possible the linkage between the high power plant and the low power microcomputer, is discussed. Then, the identified plant is presented.

Some aspects of the hardware of the 16-bit micro computer employed in the investigations are outlined in chapter 4. This chapter also explains how programs can be developed in a standard PC system and then implemented in the microcomputer used for control purposes. This implementation required initially the development of a general purpose software, that consists of subprogram libraries and a minimal real-time operating system. These facilities allow the use of vector instructions with the numeric coprocessor and the control of the input/output devices of the microcomputer system as well as the scheduling and dispatching of the controller tasks according to their specified sampling time.

In chapter 5 the basic theoretical principles of the implemented adaptive control techniques are reviewed. A direct adaptive control scheme based on an improved model reference approach and an adaptive pole placement method, which allows also minimization of a linear quadratic performance index, are presented. The discussion concentrates also on the aspects of weighted parameter estimation methods used for control algorithms with perpetual adaptation of parameters and on aspects like integrator wind-up reset, minimization of computing time etc., which are of immense practical importance.

In chapter 6 the synthesis of the controllers and the details of the realization such as initial parametrization, computation time, computer memory usage etc. are presented.

In chapter 7 the structures and the performance of the adaptive control schemes and of a conventional analog controller are compared. The results are shown in time response diagrams and the relative merits of these schemes are discussed.

Chapter 8, as conclusion, presents a review, a critical evaluation and suggestions for future investigations and extensions.

1.4 References

Balasubramanian, R. and K.H. Wong (1986), 'A microcomputer-based self tuning IP controller for DC machines', *IEEE Trans. on Ind. Appl.*, 22, pp. 989-999.

Balestrino, A., G. Maria and L. Sciavicco (1983), 'Adaptive control design in servosystems', Proc. 3rd IFAC Symp. on Control in Power Electronics and Electrical Drives, Lausanne, pp. 125-131.

Best, J. and P. Mutschler (1982), 'Methods of microcomputer based SCR-DC-motor drive control', in: W. Leonhard (Ed.), *'Microelectronics in Power Electronics and Electrical Drives'*, ETG-Fachberichte 11, VDE-Verlag, Berlin, pp. 265-271.

Bose, B.K. (1988), 'Technology trends in microcomputer control of electric machines', *IEEE Trans. on Ind. Electron.*, 35, pp. 160-177.

Brickwedde, A. (1983), 'Microprocessor-based adaptive control for electrical drives', Proc. 3rd IFAC Symp. on Control in Power Electronics and Electrical Drives, Lausanne, pp. 119-124.

Bühler, E. (1978), 'Eine zeitoptimale Thyristor-Stromregelung unter Einsatz eines Mikroprozessors', *Regelungstechnik*, 26, pp. 37-43

Buxbaum, A. (1969), 'Regelung von Stromrichterantrieben bei lückendem und nichtlückendem Ankerstrom', *Techn. Mitt. AEG-Telefunken*, 59, pp. 348-352.

Chan, Y.T, A.J. Chmiel and J.B. Plant (1980), 'A microprocessor-based current controller for SCR-DC motor drives', *IEEE Trans. on Ind. Electron. Contr. Instrum.*, 27, pp. 169-176.

Claussen, U. and G. Fromme (1978), 'Motorregelung mit Mikrorechner', *Regelungstechnische Praxis*, 20, pp. 355-359.

Collings, T.D. and W. J. Wilson (1991), 'A fast-response current controller for microprocessor-based SCR-DC motor drives', *IEEE Trans. on Ind. Appl.*, 27, pp. 921-927.

Courtiol, B. and I.D. Landau (1975), 'High speed adaptation system for controlled electrical drives', *Automatica*, 11, pp. 119-127.

De, N.K., S. Sinha and A. Chattopadhyay (1985), 'Microcomputer as a programmable controller for state feedback control of a DC motor employing thyristor amplifier', *IEEE Trans. on Ind. Appl.*, 21, pp. 571-579.

Depping, F. and M. Voits (1982), 'Microcomputer-based parameter adaptive speed control with deadbeat response algorithm for an electrical drive', In: W. Leonhard (Ed.), *'Microelectronics in Power Electronics and Electrical Drives'*, ETG-Fachberichte 11, VDE-Verlag, Berlin, pp. 265-271.

Depping, F. and M. Voits (1983), 'Automatic selection of control algorithms for an electrical drive with microcomputer speed control', Proc. 3rd IFAC Symp. on Control in Power Electronics and Electrical Drives, Lausanne, pp. 507-514.

Ernst, D and D. Ströle (1973), *'Industrieelektronik. Grundlagen-Methoden-Anwendungen'*, Springer-Verlag, Berlin.

Erzberger, H. (1968), 'Analysis and design of model following systems by state space techniques', Proc. of the Joint Automatic Control Conference, Ann Arbor, pp. 572-581.

Favre, J.P (1982), 'Microprocessor-based speed control of a SCR-DC motor', in: W. Leonhard (Ed.), *'Microelectronics in Power Electronics and Electrical Drives'*, ETG-Fachberichte 11, VDE-Verlag, Berlin.

Fromme, G. (1982), 'Self-optimising controller employing microprocessors for plant with slowly or discontinuously varying parameters', *Process Automation*, R. Oldenbourg Verlag, München, pp. 93-99.

Golde, E. and K.H Riebschläger (1971), 'Stromregelung für kreisstromfreie Stromrichterschaltungen', *Techn. Mitt. AEG-Telefunken*, 61, pp. 135-137.

Grötzbach, M. (1981), 'Diskretes Kleinsignalverhalten pulsbreitengesteuerter Stromrichter im Lückbetrieb', ETZ Archiv, 3, pp. 91-93.

Grotstollen, H. (1977), 'Comparison of speed controlled DC drives with and without subordinate current loop', Proc. 2nd IFAC Symp. on Control in Power Electronics and Electrical Drives, Düsseldorf, pp. 583-592.

Hahn, V. and H. Unbehauen (1982), 'Direct adaptive control of nonminimum phase systems', Prepr. IEEE Conf. on Applications of Adaptive and Multivariable Control, Hull, pp. 170-175.

Hahn, V., H. Unbehauen and U. Nadolph (1983), 'Model reference adaptive control of a multivariable nonminimum phase pilot plant', Prepr. of the 3rd Yale Workshop on Applications of Adaptive System Theory, Yale, pp. 41-46.

Hahn, V. (1983), 'Direct adaptive control schemes for discrete time control of multivariable systems', (in German), Dr.-Ing. Thesis, Ruhr-Universität Bochum.

Hahn, V. (1985), 'A direct adaptive controller for nonminimum phase systems', Proc. 7th IFAC/IFIP/IMACS Conference on Digital Computer Applications to Process Control, Wien, pp. 523-528.

Harashima, F. (1983), ' Speed control of motor drives via microprocessor', In: S.G. Tzafestas (ed.), *'Microprocessors in signal processing, measurement and control'*, D. Reider Publishing Company, Dordrecht/Boston/Lancester, pp. 283-298.

Hasegawa, T., T. Nakagawa, H. Hosoda, R. Kurasawa and H. Naitoh (1984), 'A microcomputer-based thyristor Leonhard system having powerful RAS functions', *IEEE Trans. on Ind. Electron.*, 31, pp. 74-78.

Hill, R.J. and F.L.Luo (1987), 'Stability analysis of thysistor current controllers', *IEEE Trans. on Ind. Appl.*, 23, pp 49-56.

Holtz, J. and V. Schwellenberg (1983), 'A new fast-response current control scheme for line controlled converters', *IEEE Trans. on Ind. Appl.*, 19, pp. 579-585.

Hong, Z.D. and M.A. Zohdy (1985), 'An adaptive control scheme for oscillatory servo systems', *IEEE Trans. on Ind. Electron.*, 32, pp. 37-40.

Hsu,Y.Y and W.C. Chan (1984), 'Optimal variable-structure controller for DC-motor speed control', *IEE Proc. Control Theory and Applications*, 131, pp. 233-237.

Joos, G. and T.H. Barton (1975), 'Four quadrant DC-variable-speed drives', *Proceedings of the IEEE*, 63, pp. 1660-1668.

Joos, G. and E.D. Goodman (1983), 'An adaptive controller for DC-drives in discontinuous current mode', IEEE International Electrical, Electronics Conference and Exposition, Toronto, pp. 26-28.

Joos, G., P. Sicard and E. Goodman (1992), 'A comparison of microcomputer-based implementations of cascaded and parallel speed and current loops in DC motor drives', *IEEE Trans. on Ind. Appl.*, 28, pp. 136-143.

Kelly, R. (1987), 'A linear state feedback plus adaptive feed-forward control for DC servomotors', *IEEE Trans. on Ind. Electron.*, 34, pp. 153-157.

Kennel, R. and D. Schröder (1983), 'Predictive control strategy for converters', 3rd IFAC Symp. on Control in Power Electronics and Electrical Drives, Lausanne, pp. 415-422.

Kümmel, F. (1965), 'Einfluß der Stellgliedeigenschaften auf die Dynamik von Regelkreisen mit unterlagerter Stromregelung', *Regelungstechnik*, 13, pp. 227-234.

Landau, I.D. (1979), *'Adaptive Control: The Model Reference Approach'*, Marcel Dekker Inc., NewYork/Basel.

Leonhard, W. (1980), 'Mikrorechner in der elektrischen Antriebstechnik', in: D. Ernst und M. Thoma (Eds.), *Meß- und Automatisierungstechnik*, Fachberichte Interkama-Kongreß, 5, Springer-Verlag, Berlin, pp. 637-655.

Lozano, R. and I.D. Landau (1981), 'Redesign of adaptive control schemes', *Int. J. Contr.*, 33, pp. 247-268.

Lozano, R. and A. Noriega (1983), 'Microcomputer implementation of an adaptive control algorithm', Prepr. IFAC/IFIP Symposium in Real-Time Digital Control Applications, Mexico, pp. 108-113.

Magyar, P., E. Schnieder and W. Vollstedt (1982), 'Digitale Regelung und Steuerung einer stromrichtergespeisten Gleichstrommaschine mit Mikrorechner', *Regelungstechnik*, 30, pp. 378-387.

McLaren, S.G., M.G. Rodd, A.P. von Zwiklitz and H.F. Weehuizen (1975), 'Direct digital control of thyristor drives', Proc. 1st IFAC Symp. on Control in Power Electronics and Electrical Drives, Düsseldorf, pp. 146-159.

Naitoh, H. and S. Tadakuma (1987), 'Microprocessor-based adjustable-speed DC motor drives using model reference adaptive control', *IEEE Trans. on Ind. Appl.*, 23, pp. 313-318.

Nandam, P.K. and P.C. Sen (1987), 'Analog and digital speed control of DC drives using proportional-integral and integral-proportional control techniques', *IEEE Trans. on Ind. Electron.*, 34, pp. 227-233.

Ohishi, K., K. Ohnishi and K. Migachi (1988), 'Adaptive DC servo drive control taking force disturbance suppression into account', *IEEE Trans. on Ind. Appl.*, 24, pp. 171-176.

Ohmal, T., T. Matsuda, T. Suzuki, N. Azusawa, K. Kamiyama and T. Konishi (1980), 'A microprocessor-controlled fast-response speed regulator with dual-mode current loop for DCM drives', *IEEE Trans. on Ind. Appl.*, 16, pp. 388-394.

Ohmal, T., T. Matsuda, R. Masaki, K. Saito (1986), 'A microprocessor-based current controller with an internal current-rate loop for motor drives', *IEEE Trans. on Ind. Appl.*, 22, pp. 805-811.

Oumar, A., J.P. Louis and A. El-Hefnawy (1977), 'Design of an optimal autoadaptive current loop for DC motor', 2nd IFAC Symp. on Control in Power Electronics and Electrical Drives, Düsseldorf, pp. 593-601.

Parks, P.C., W. Schaufelberger, Chr. Schmid and H. Unbehauen (1980), 'Applications of adaptive control systems', in: H. Unbehauen (Ed.), *Methods and Applications in Adaptive Control*, Springer Verlag, pp. 161-198.

Pelly, B.R. (1971), *'Thyristor Phase-Controlled Converters and Cycloconverters'*, John Wiley & Sons, New York/London.

Platzer, D. and H. Kaufman (1984), 'Model reference adaptive control of a thyristor driven DC motor system subject of current limitations', Prepr. 9th IFAC World Congress, Budapest, pp. 231-235.

Raatz, E. (1970), Der Einsatz von adaptiven Drehzahlreglern in der Antriebstechnik', *Techn. Mitt. AEG-Telefunken*, 60, pp. 375-378.

Riedo, P.J. (1982), 'Cascade digital control by state-variable feedback method applied to a DC-motor', in: W. Leonhard (Ed.), *'Microelectronics in Power Electronics and Electrical Drives'*, ETG-Fachberichte 11, VDE-Verlag, Berlin.

Schnieder, E. (1977), 'Control of DC-drives by microprocessors', Proc. 2nd IFAC Symp. on Control in Power Electronics and Electrical Drives, Düsseldorf, pp. 603-608.

Sen, P.C. (1990), 'Electric motor drives and control - Past, present and future', *IEEE Trans. on Ind. Electron.*, 37, pp. 562-575.

Shirakura, M., M. Ohara, K. Ishida and K. Ueto (1983), 'Direct digital controlled thyristor Leonhard with a state observer', 9th Annual Conference on Industrial Electronics, San Francisco, pp. 70-74.

Sinner, E. (1973), 'Regulateur adaptif a variables d'etat pour processus industriels', VDE Deutsch-Französischer Aussprachetag, Industrielle Anwendung adaptiver Systeme, Freiburg, pp. 115-136.

Slattery, D.T. and P.A. Wade (1983), 'Digital trends in the design and application of DC-drives', Proc. of the Conference on Drives/Motors/Controls '83, Harrogate, pp. 62-68.

Sobel, K., H. Kaufman and L. Mabius (1982), 'Implicit adaptive control for a class of MIMO systems', *IEEE Transactions on Aerospace and Electronic Systems*, 18, pp- 576-590.

Ströle, D. (1967a), 'Adaptivsysteme der elektrischen Antriebstechnik', *ETZ-Archiv*, 88, pp. 182-185.

Ströle, D. (1967b), 'Typische Adaptivsteuerung bei geregelten elektrischen Antrieben', *Regelungstechnik*, 15, pp. 106-111.

Tadakuma, S., Y. Tamura and M. Hirano (1987), 'Finite-time settling control based current controller for thyristor dual converter', *IEEE Trans. on Ind. Appl.*, 23, pp. 603-609.

Tamura, K., K. Ogata and P.N. Nikiforak (1991), 'Design of nonovershoot MRACS with application to DC servo motor system', *ASME Journal of Dynamic Systems, Measurement and Control*, 113, pp. 75-81.

Unbehauen, H. (1985), *'Regelungstechnik III: Identifikation, Adaption, Optimierung'*, Vieweg Verlag, Braunschweig/Wiesbaden.

Utkin, V.I. (1993), 'Sliding mode control design principles and applications to electric drives', *IEEE Trans. on Ind. Electron.*, 40, pp. 23-36.

Weihrich, G. (1978), 'Drehzahlregelung von Gleichstromantrieben unter Verwendung eines Zustands- und Störgrößenbeobachters', *Regelungstechnik*, 26, pp. 349-380 and 392-397.

Weihrich, G. and D. Wohld (1980), 'Adaptive speed control of DC-drives using adaptive observers', *Siemens Forschung und Entwicklung*, 9, pp. 283-287.

Wiemer , P., V. Hahn, Chr. Schmid and H. Unbehauen (1983), 'Application of multivariable model reference adaptive control to a binary distillation column', Prepr. of the IFAC Workshop on Adaptive Systems in Control and Signal Processing, San Francisco, paper AAC-1.

CHAPTER 2
A SIMPLE MODEL FOR A THYRISTOR DRIVEN DC-MOTOR CONSIDERING CONTINUOUS AND DISCONTINUOUS CURRENT MODES

2.1 Introduction

Modelling is an indispensable step in the synthesis of high performance control systems. The model must represent the most relevant characteristics of the system for the proposed application. The modelling of a DC-motor is by now standard and discussed in traditional text-books (Slemon, 1966; Fitzgerald et. al, 1971; Fröhr and Ortenburger, 1971). However, when the DC-motor is supplied with an AC-DC converter, the current and armature voltage in steady state are no longer smooth. They are composed of a mean-value component and sinusoidal harmonic terms. These terms influence the construction of the motor (Robinson, 1968), as well as the "mean-value" or "pseudo-instantaneous" (Bland, 1967) model, which will be discussed later, due to the existence of a discontinuous current mode of operation. The proposed model (Stepahn, 1991) will be established for a single-phase bridge, as in this case the influence of the bridge on the DC-motor performance is larger than in three or six phase bridges. Nevertheless, the motor model can be easily extended to those configurations.

Fig. 2.1 shows a fully controlled single-phase thyristor bridge supplying a DC-motor. The AC input voltage is given by $v=\sqrt{2}U_s\omega t$. The well known (Buxbaum, 1980; Pfaff, 1982) steady-state relationship between the firing angle α, the current mean-value I_d and the back voltage e, for negligible source impedance (Z=0), is repeated in Fig. 2.2. These curves indicate that the relation between the input voltage and current is nonlinear in the discontinuous current mode.

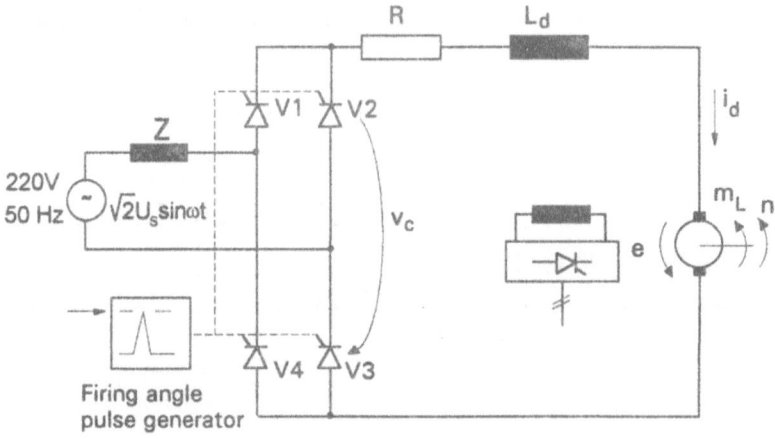

Fig. 2.1 DC-motor with thyristor power supply

The dynamic variation of the current for a step variation of the firing angle from α_1 to α_2 in discontinuous current operation mode is shown in Fig. 2.3a. It can be seen that the current reaches its steady-state wave form during the first cycle after the new firing angle comes into operation. This characteristic indicates that the armature time constant in the discontinuous mode has little influence on a mean-value model. It is important to notice that the armature time constant influences the shape of conduction periods, but the current "mean-value" changes as a system without time constants.

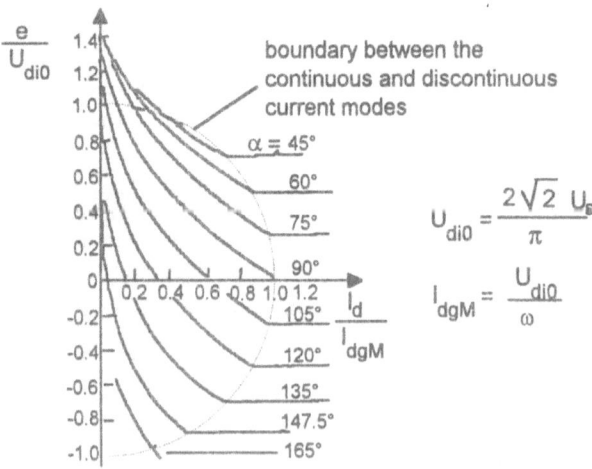

Fig. 2.2 Steady-state characteristics of a DC-motor fed by a single-phase converter

14

On the other hand, in continuous current mode, there is a current "mean-value" dynamics, as is shown in Fig. 2.3b. These remarks were made by some authors and were used to develop heuristic "mean-value" models for the thyristor fed DC-motor (Buxbaum, 1980; Pfaff, 1982). A mathematical explanation of this model will be given in the following sections.

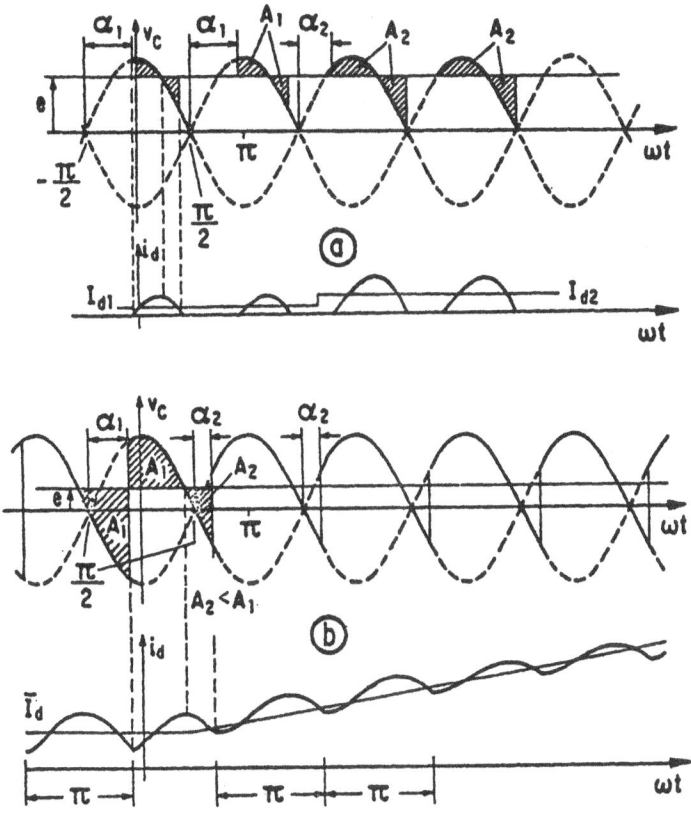

Fig. 2.3 Variation of armature current for a firing angle variation from α_1 to α_2.
(a) discontinuous current mode   (b) continuous current mode

2.2 The DC-Motor Model and the Basic Equations

The armature circuit of a DC-motor is represented by the back voltage e, the armature inductance L_d and the armature resistance R. If the supply voltage is v_c and i_d is the current, it follows that

15

$$v_c(t) - e(t) = R\, i_d(t) + L_d \frac{di_d(t)}{dt}.$$ (2.1)

The back emf e is related to the speed n by

$$e(t) = k_1\, \psi(t)\, n(t),$$ (2.2)

where k_1 is a machine constant and ψ the magnetic flux. The developed torque is given by

$$m(t) = k_2\, \psi(t)\, i_d(t),$$ (2.3)

where k_2 is a machine constant. The torque balance equation is

$$m(t) - m_L(t) = B'\, n(t) + 2\pi\theta \frac{dn(t)}{dt}$$ (2.4)

where θ is the inertia, m_L the load torque and B' the friction coefficient.

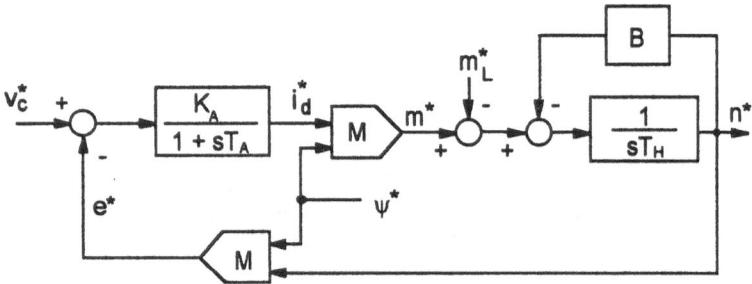

Fig. 2.4 Block diagram of a separately excited DC-motor

The normalized equations are

$$v_c^*(t) - e^*(t) = \frac{1}{K_A} [\, i_d^* + T_A \frac{di_d^*(t)}{dt}\,],$$ (2.5)

$$e^*(t) = \psi^*(t)\, n^*(t),$$ (2.6)

$$m^*(t) = \psi^*(t)\, i_d^*(t),$$ (2.7)

$$m^*(t) - m_L^*(t) = B\, n^*(t) + T_H \frac{dn^*(t)}{dt},$$ (2.8)

with

$$v_c^* = \frac{v_c}{U_N}, \quad e^* = \frac{e}{U_N}, \quad i_d^* = \frac{i_d}{I_N}, \quad \psi^* = \frac{\psi}{\phi_N}, \quad n^* = \frac{n}{N_0}, \quad m^* = \frac{m}{M_N},$$

$$B = B' \frac{M_0}{M_N}, \quad T_H = \frac{2\pi\theta N_0}{M_N}, \quad K_A = \frac{U_N}{RI_N}, \quad T_A = \frac{L_d}{R}, \tag{2.9}$$

where

U_N	is the voltage reference value,
I_N	is the current reference value,
ϕ_N	is the flux reference value,
$M_N = k_2\phi_N I_N$	is the torque reference, and
$N_0 = \dfrac{U_N}{k_1\phi_N}$	is the speed reference value.

This model can be presented as a block diagram such as in Fig. 2.4.

2.3 The Model of a Thyristor Driven DC-Motor

2.3.1 Armature Model

Suppose f(t) is a time function. The following "mean-value" function will be defined

$$\bar{f}(t) \triangleq \frac{\omega}{\pi} \int_{t-\frac{\pi}{2\omega}}^{t+\frac{\pi}{2\omega}} f(\tau)\, d\tau . \tag{2.10}$$

In steady-state conditions, this definition is equivalent to the common mean-value for periodic functions, i.e.

$$\bar{f}(t) = F = \frac{\omega}{\pi} \int_{0}^{\frac{\pi}{\omega}} f(\tau)\, d\tau . \tag{2.11}$$

Differentiation of Eq. 2.10 gives

$$\frac{d\bar{f}(t)}{dt} = \frac{\omega}{\pi}\, [\, f(t+\frac{\pi}{2\omega}) - f(t-\frac{\pi}{2\omega})] = \frac{\omega}{\pi} \int_{t-\frac{\pi}{2\omega}}^{t+\frac{\pi}{2\omega}} \frac{df(\tau)}{d\tau}\, d\tau. \tag{2.12}$$

Applying definition (2.10) in Eq. (2.5), it follows that

17

$$\frac{\omega}{\pi} \int_{t-\frac{\pi}{2\omega}}^{t+\frac{\pi}{2\omega}} (v_c^*(\tau) - e^*(\tau))d\tau =$$

$$\frac{1}{K_A} \left[\frac{\omega}{\pi} \int_{t-\frac{\pi}{2\omega}}^{t+\frac{\pi}{2\omega}} i_d^*(\tau)d\tau + \frac{\omega}{\pi} \int_{t-\frac{\pi}{2\omega}}^{t+\frac{\pi}{2\omega}} T_A \frac{di_d^*(\tau)}{d\tau} d\tau \right].$$ (2.13)

Using Eqs. (2.10) and (2.12) in Eq. (2.13) one obtains

$$(\overline{v}_c^*(\tau) - \overline{e}^*(\tau)) = \frac{1}{K_A} \left[\overline{i}_d^*(t) + T_A \frac{d\overline{i}_d^*(t)}{d\tau} \right],$$ (2.14)

where $\overline{v}_c^*$, $\overline{e}^*$, and $\overline{i}_d^*$ are respectively the "mean-value" functions of the armature voltage v_c^*, back voltage e^* and current i_d^* defined in Eq. (2.10).

In view of the negligible variation of the back voltage during the integration interval $T=\pi/\omega$ (see Eq. (2.10)), it follows that

$$\overline{e}^*(t) = e^*(t).$$ (2.15)

2.3.1.1 Discontinuous Current Mode
In discontinuous current mode $\overline{i}_d^*(t)$ can be considered constant as shown in Fig. 2.3(a), therefore

$$\frac{d\overline{i}_d^*(t)}{dt} = 0.$$

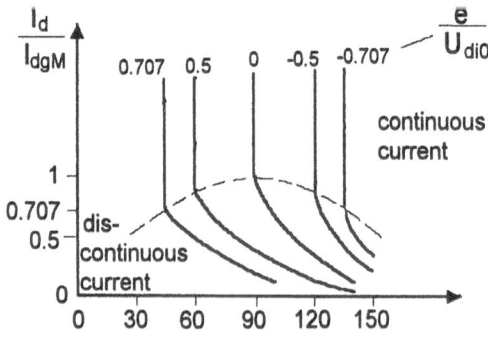

Fig. 2.5 A different representation of Fig. 2.2

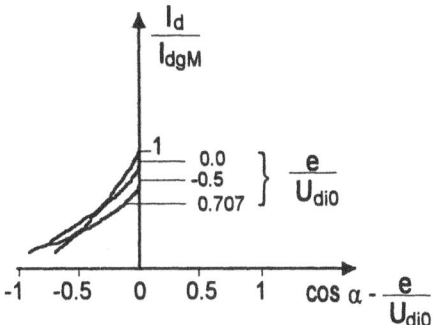

Fig. 2.6 Another representation of Fig. 2.5 in the discontinuous current mode.

Moreover, the relation between the firing angle α, the current mean-value I_d and the back emf e shown in Fig. 2.2 may also be given as shown in Fig. 2.5 (Pfaff, 1982). If the coordinate of the abscissa is then conveniently altered to $\cos\alpha - (e/U_{dio})$, one obtains Fig. 2.6. The block diagram in Fig. 2.7 is a consequence of Fig. 2.6 conveniently scaled. One can now easily see that the converter-armature gain is nonlinear in the discontinuous current "mean-value" model.

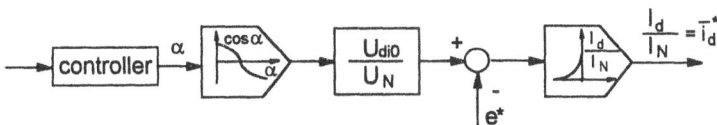

Fig. 2.7 Rectifier/motor armature "mean-value" model (discontinuous current)

2.3.1.2 Continuous Current Mode

In the continuous current mode, it is possible to use the approximation

$$\vec{v}_c^*(t) = \frac{U_{dio}}{U_N} \cos \alpha(t) \,.$$ (2.16)

Eq. (2.14) can then be rewritten as

$$\frac{U_{dio}}{U_N} \cos \alpha(t) - e^*(t) = \frac{1}{K_A} [\, \vec{i}_d^*(t) + T_A \frac{d\vec{i}_d^*(t)}{dt} \,].$$ (2.17)

The same result is shown as a block diagram in Fig. 2.8.

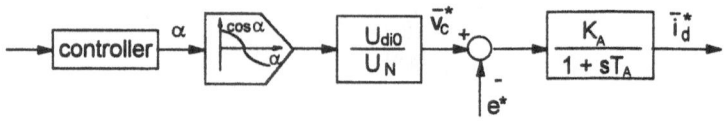

Fig. 2.8 Rectifier/motor armature "mean-value" model (continuous current)

2.3.2 Complete Mean-Value Model

The application of definition (2.10) in Eq. (2.7) gives

$$\bar{m}^*(t) = \frac{\omega}{\pi} \int_{t-\frac{\pi}{2\omega}}^{t+\frac{\pi}{2\omega}} m^*(\tau)d\tau = \frac{\omega}{\pi} \int_{t-\frac{\pi}{2\omega}}^{t+\frac{\pi}{2\omega}} \psi^*(\tau)\, i_d^*(\tau)d\tau . \tag{2.18}$$

In view of the great field time constant, the variation of the field excitation can be neglected during an integration interval of $T = \pi/\omega$. It follows that

$$\bar{m}^*(t) = \psi^*(t)\, \bar{i}_d^*(t). \tag{2.19}$$

The same consideration is valid for Eq. (2.6), therefore

$$\bar{e}^*(t) = \psi^*(t)\, \bar{n}^*(t). \tag{2.20}$$

Considering the fact that the speed $n^*(t)$ cannot vary significantly in a time interval of $T = \pi/\omega$, and using Eq. (2.15) in Eq. (2.20) one can write

$$e^*(t) = \psi^*(t)\, n^*(t). \tag{2.21}$$

Applying the "mean-value" definition of Eq. (2.10) to Eq. (2.8), one obtains

$$\bar{m}^*(t) - \bar{m}_L^*(t) = B\, n^*(t) + T_H \frac{dn^*}{dt} . \tag{2.22}$$

The results summarized in Figs. 2.7 and 2.8, as well as in Eqs. (2.19), (2.21) and (2.22) lead to the block diagram shown in Fig. 2.9. The signals within this non-linear block diagram are considered to be in the time domain. A variable dead time T_t was introduced to represent more properly the rectifier (Schröder, 1972; Lagasse, 1974). K_A and T_A are constants in the continuous current mode. In discontinuous current mode K_A is nonlinear, as shown in Fig. 2.6, and $T_A = 0$.

This resulting model is such that the structure remains the same in both current conduction modes. The difference lies exclusively in variations of the armature gain and time constant. The model includes all significant aspects necessary to design simple controllers.

The comparison of Fig. 2.4 with Fig. 2.9 shows that the converter introduces in the DC-motor model:

- a variable dead time,

- a nonlinear cosine function,

- a nonlinear armature performance, depending on the current conduction mode (this nonlinearity is usually compensated by a dual-mode adaptive armature current controller).

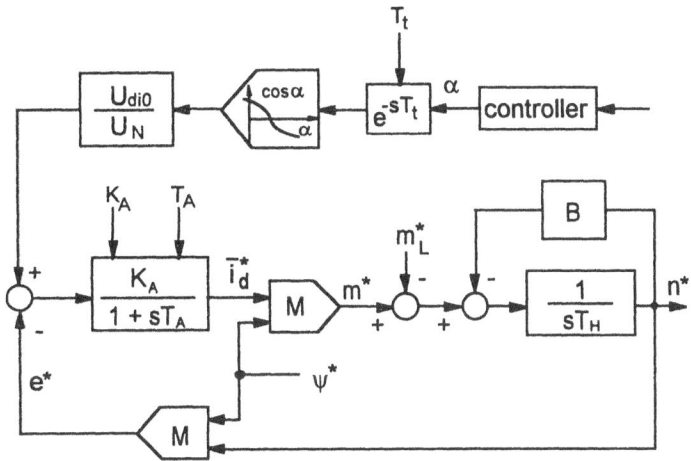

Fig. 2.9 Complete "mean-value" model of a converter fed DC-motor.

Experimental tests can be carried out to obtain the curves presented in Fig. 2.6 and thus determine the nonlinear gain K_A in the discontinuous conduction mode. The parameters K_A and T_A, in the continuous conduction mode, are easily obtained with step response methods.

2.4 Experimental Results

As was shown in section 2.2, the converter influences principally the model of the armature. Therefore, in this section, the study of a current control loop, where the electro-mechanical behaviour is of secondary importance, will be presented.

The experimental system consists of a thyristor fed 1.1 kW separately excited DC-machine. The armature power control rectifier is a dual-converter containing two thyristor bridges, each one with four thyristors in fully controlled single-phase configuration. The current controller will be dual-mode adaptive: in continuous current mode it is a PI-controller, in discontinuous current mode it is an I-controller with variable gain (Buxbaum, 1980). The control loop is depicted in Fig. 2.10 and the motor back emf will be regarded as a disturbance. The identified parameters are also shown in this figure.

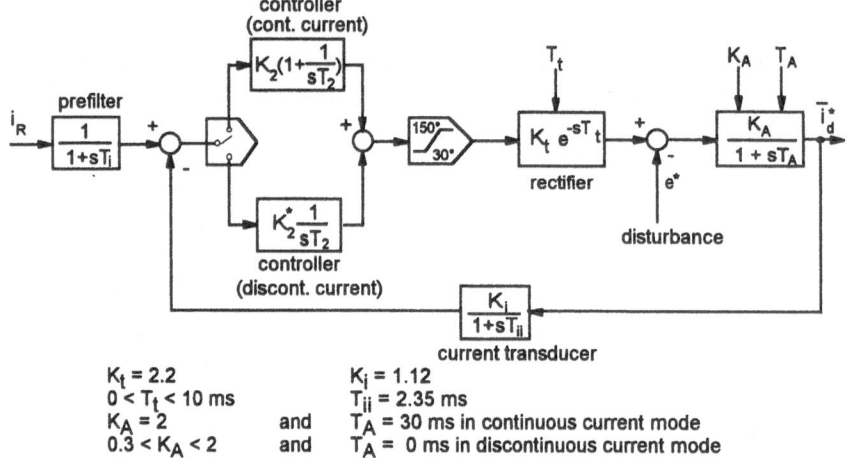

$K_t = 2.2$ $K_i = 1.12$
$0 < T_t < 10$ ms $T_{ii} = 2.35$ ms
$K_A = 2$ and $T_A = 30$ ms in continuous current mode
$0.3 < K_A < 2$ and $T_A = 0$ ms in discontinuous current mode

Fig. 2.10 The analog dual-mode adaptive current loop.

Representing the rectifier variable dead time ($0<T_t<10$ms) by a statistical mean-value T_{to} of 5ms (Lagasse et al., 1987) and applying Kessler's "Betrags-Optimum" method (Kessler, 1955; Umland and Safuddin, 1990), the optimal controller parameters for continuous current are determined as

$$T_2 = T_A = 30 \text{ ms}, \tag{2.23}$$

$$K_2 = \frac{T_A}{2K_t K_A K_i (T_{to} + T_{ii})} = 0.41, \tag{2.24}$$

$$T_i = T_{ii} = 2.35 \text{ ms}. \tag{2.25}$$

For the discontinuous current operating mode, the I-controller must be such that the linearized open-loop transfer function remains equal to the open-loop transfer function in continuous current operating mode. Then

$$K_2^* \frac{1}{sT_2} K_{A(disc.)} = K_2 \left(1 + \frac{1}{sT_2}\right) \frac{K_{A(cont.)}}{1 + sT_A} \tag{2.26}$$

where $0.3 < K_{A(disc.)} < 2$ represents the nonlinear gain K_A in discontinuous current mode, and $K_{A(cont.)} = 2$ represents the gain K_A in continuous current mode. Using Eq. (2.23), it follows

$$K_2^* = K_2 \frac{K_{A(cont.)}}{K_{A(disc.)}} . \tag{2.27}$$

The parameters were adjusted in a commercially available analog controller with resistors and capacitors.

Experimental results showing step responses of the current control loop at different operating conditions are presented in Fig. 2.11. The mean-value obtained using Eq. (2.10) is also depicted in this figure.

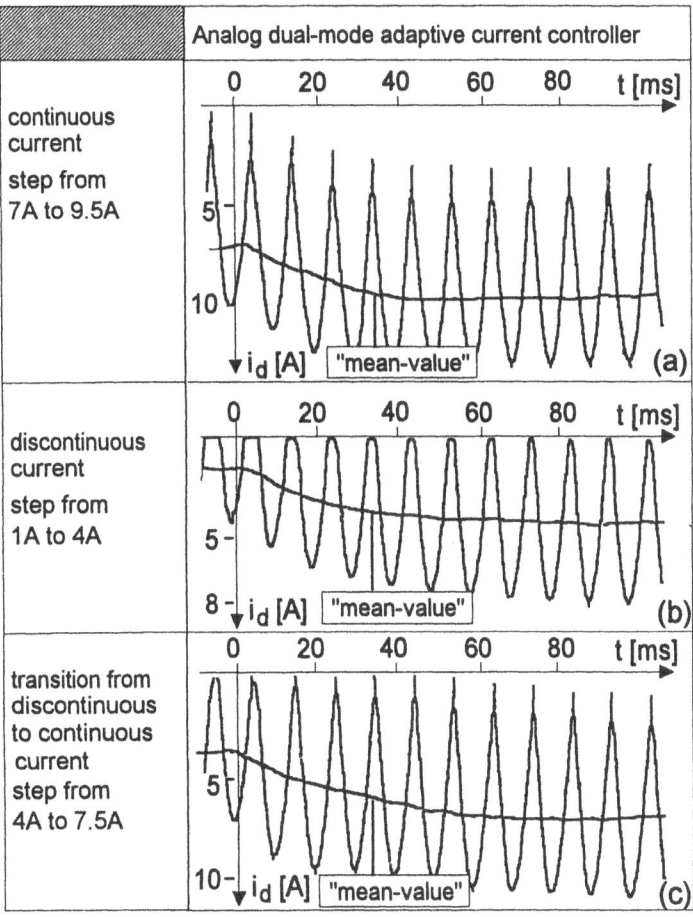

Fig. 2.11 Step responses of the current loop at different operating conditions.

The simulation of the conditions of continuous or discontinuous current, using the "mean-value" model, is presented in Fig. 2.12. As the adaptive current controller compensates for the plant variations in order to give the same transfer function in continuous or discontinuous current modes, the simulation is theoretically the same in both cases. Moreover, as the simulated closed loop system is linear and time invariant, it is sufficient to observe the unit step response to know the system behaviour.

The "mean-value" for continuous current presented in Fig. 2.11(a) and the simulation of Fig. 2.12 show the same dead-time and practically the same rise and settling time. Figs. 2.11(b) and (c) don't match exactly with Fig. 2.12 because the practical compensation of the dual-mode adaptive controller is an approximation of the ideal condition given by Eq. (2.27).

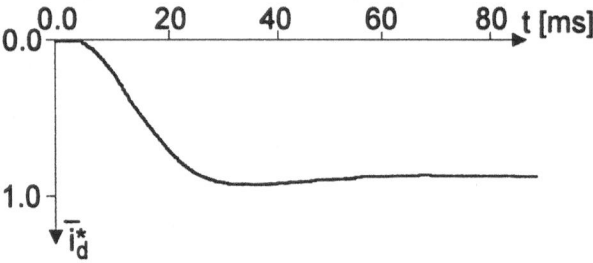

Fig. 2.12 Simulation result of the unit step response of the current control loop.

Nevertheless, the simulated result and the "mean-value" obtained by experimental results are similar and validate the proposed approach.

The same model was also used to project a robust current controller and the results were reported in another publication (Unbehauen et al., 1987).

2.5 Conclusion

A mathematical explanation for a "mean-value" or "pseudo-instantaneous" model of a thyristor fed DC-motor has been presented. The proposed model has the same structure for continuous and discontinuous current condition modes and includes all significant aspects necessary to design simple controllers. The model was used for the synthesis of a dual-mode adaptive current controller. The comparison of experimental and simulation results validated the proposed model.

2.5 References

Bland, R. J. (1967), 'Factors affecting the operation of phase-controlled cycloconverter', *Proc. IEEE*, 114, pp. 1908-1916.

Buxbaum, A. and K. Schierau (1980), *'Berechnung von Regelkreisen der Antriebstechnik'*, AEG-Telefunken, Frankfurt.

Fitzgerald, A. E., C. Kingsley and A. Kusko (1971), *'Electric Machinery'*, McGraw Hill, New York.

Fröhr, F. and F. Ortenburger (1971), *'Technische Regelstreckenglieder bei Gleichstromantrieben'*, Siemens, Berlin/München.

Kessler, C., (1955), 'Über die Vorausberechnung optimal abgestimmter Regelkreise', *Regelungstechnik*, Vol. 3, pp. 40-48.

Lagasse, J. and R. Prajoux (1974), 'Behaviour of control systems including controlled convertors, especially rectifiers: A review of existing theories', Proc. 1st IFAC Symp. Contr. in Power Electron. Elect. Drives, Düsseldorf, pp. 1-37.

Pfaff, G. and C. Meier (1982), *'Regelung elektrischer Antriebe II'*, Oldenbourg Verlag, München.

Robinson, C. E. (1968), 'Redesign of DC motors for applications with thyristor power supplies', *IEEE Trans. Industry General Appl.*, 4, pp. 508-514.

Schröder, D. (1972), 'Analysis and synthesis of automatic control systems with controlled converters', Proc. IFAC 5th World Congress on Automat. Contr., Paris, paper P22.1.

Slemon, G. R. (1966), *'Magnetoelectric Devices'*, Wiley, New York.

Stephan, R.M. (1991), 'A simple model for a thyristor driven DC motor considering continuous and discontinuous current modes', IEEE Trans. on Education, 34, pp. 330-335.

Umland, J.W. and M. Safuddin (1990), 'Magnitude and Symmetric Optimum criterion for the design of linear control systems: What is it and how does it compare with the others?', IEEE Trans. on Ind. Appl., 26, pp. 489-497.

Unbehauen, H., J. Dastych and R. Stephan (1987), 'Robust current control for a thyristor driven DC-motor', Proc. 10th IFAC World Congress on Automat. Contr., München, pp. 331-336.

CHAPTER 3
THE EXPERIMENTAL LABORATORY SYSTEM

3.1 Introduction

The experimental laboratory system is an improvement of the Siemens' "Teaching and Training Demonstration Model for Variable Speed Drive" (Siemens, 1977). The plant to be controlled consists of a DC-motor fed by a single-phase fully-controlled dual-converter. Variations of the moment of inertia, field excitation and load torque are easily accomplished with potentiometers or switches. A current transducer, a speed transducer, as well as voltage, current, speed and torque measurement instruments are available. Several connection jacks make the observation of various internal signals possible. An interface unit allows the connection of the motor to a microcomputer, where a digital controller can be programmed. An analog controller is also available. Control keys on an operating console make the substitution between analog and digital controllers an easy task. This laboratory system will be described in the following sections. The model parameters will also be presented.

3.2 Construction

The equipment (Fig. 3.1) comprises a fundamental unit and an extension unit (Siemens, 1977; Wieser and Stephan, 1984).

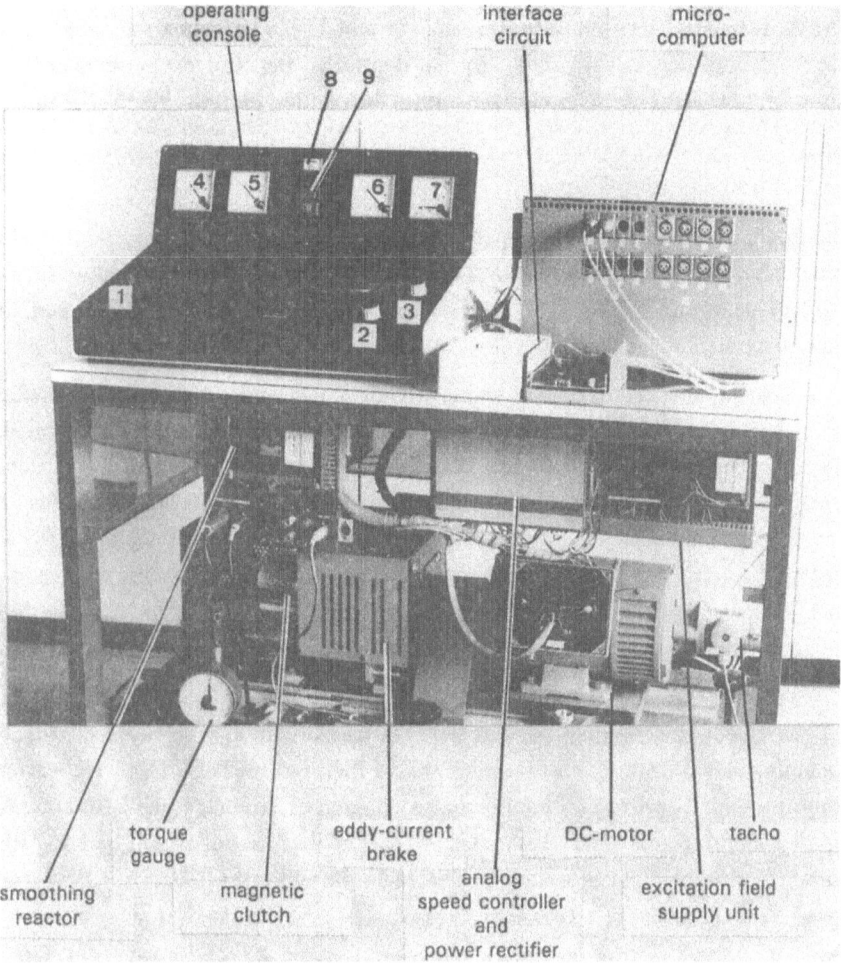

(1) set point potentiometer for speed
(2) set point potentiometer for the eddy-current brake
(3) set point potentiometer for field excitation
(4) measurement of the reference and actual speeds
(5) measurement of the armature current
(6) measurement of the armature voltage
(7) measurement of the field and brake currents
(8) clutch on-off command
(9) switches for on-off

Fig. 3.1 The experimental laboratory system

3.2.1 The Fundamental Unit

The fundamental unit consists of a 1.1 kW DC-machine, tachogenerator, eddy-current brake and its supply unit, magnetic clutch and torque gauge. The eddy-current supply unit is a single-phase half-controlled thyristor converter with a PI current controller. The DC-motor constitutes the plant to be controlled.

3.2.2 The Extension Unit

The extension unit contains the operating console, a commercially available analog speed controller, the armature power rectifier, the field excitation supply unit, the current smoothing reactor and the interface circuit between the motor and the microcomputer.

A schematic diagram of the demonstration model and setpoint potentiometers for speed (1), eddy-current brake (2) and field excitation (3) can be seen on the operating console. Instruments are available to measure the reference and actual speeds (4), armature current (5), armature voltage (6), field and brake currents (7). The key-operated switches for on-off (9) and for the clutch on-off command (8) are located between the measurement instruments. The clutch permits the sudden change of the plant moment of inertia. On the right-hand side of the console there are jacks for internal measurements and switches for commutation between the analog and the digital control. The fuses are located at the back of the rigid case.

The analog speed controller has the classical structure, whereby a dual-mode adaptive inner current loop is cascaded with a PI-speed regulator loop. The current controller switches from a PI-structure, in continuous current, to an I-structure, in discontinuous current. The performance of this analog controller will be compared with that of the digital controllers developed by the authors.

The armature power controlled rectifier contains two thyristor bridges, each one with four thyristors in fully controlled single-phase configuration. This is the plant actuator and it permits the operation in four quadrants.

The field excitation supply unit is a single-phase half-controlled thyristor converter with a PI current controller. It permits the variation of the motor field intensity and, therefore, can be used to change a parameter of the plant.

The interface circuit makes the necessary link between the motor and the microcomputer. The principal functions of this unit, which was designed and constructed during this investigation, are:

- Voltage supply for the necessary integrated circuits.

- Line voltage crossover recognition. This signal is used to synchronize the thyristor firing pulses with the line frequency.

- Current crossover recognition. This signal indicates the current flow and it is used in the dual-mode adaptive current controller. Moreover, this signal is necessary to command a thyristor-bridge change.

- Current mean-value calculation during a half-period (10 ms). The digital controller measures the current through a 12 bit A/D converter at every sampling step. Owing to the oscillatory current wave form, a direct measurement without previous calculation of the mean-value would give incorrect values for the evaluation of the current controllers.

- Generation of an impulse train of 10 kHz frequency necessary to guarantee a secure thyristor firing.

- Generation of the firing pulse signals. The firing angle information given by the microcomputer must be processed to produce a gate pulse with sufficient energy and duration.

This interface has been described in details in (Harland, 1985; Wieser, 1985).

3.3 The System Parameters

The plant was identified by means of the method suggested by Langhoff and Raatz (1976) and Bystron (1979). The mean value model used has been already presented in Fig. 2.9. The identified parameters and their variations are summarized in Fig. 3.2. The variations in the electro-mechanical time constant T_H are achieved with the magnetic clutch, which permits the connection of different masses to the motor axis. The field current can be varied and therefore the field excitation ψ is not constant. The load torque $\bar{m}_L$ and the friction coefficient B can be altered, controlling the excitation current i_b of the eddy-current brake as shown in Fig. 3.3. The nonlinear cosine relationship between firing angle and armature voltage can be compensated with a self-compensating firing circuit, also known as inverse cosine control (Groenenboom, 1972). In this study, a linearized gain K_t, as suggested by Bystron in section 3.10.6 of his book (1979), was used.

The technical data are presented in Table 3.1.

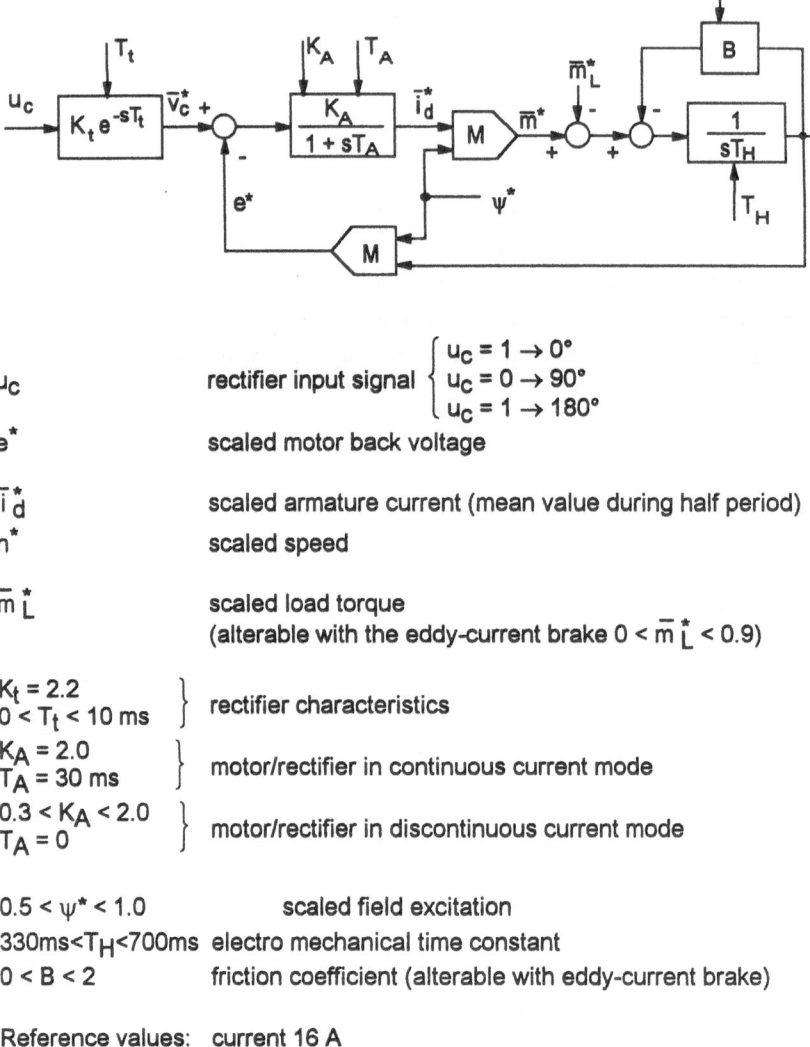

| u_c | rectifier input signal | $\begin{cases} u_c = 1 \rightarrow 0° \\ u_c = 0 \rightarrow 90° \\ u_c = 1 \rightarrow 180° \end{cases}$ |

e^* — scaled motor back voltage

$\bar{i}_d^*$ — scaled armature current (mean value during half period)

n^* — scaled speed

$\bar{m}_L^*$ — scaled load torque
(alterable with the eddy-current brake $0 < \bar{m}_L^* < 0.9$)

$\left.\begin{array}{l} K_t = 2.2 \\ 0 < T_t < 10 \text{ ms} \end{array}\right\}$ rectifier characteristics

$\left.\begin{array}{l} K_A = 2.0 \\ T_A = 30 \text{ ms} \end{array}\right\}$ motor/rectifier in continuous current mode

$\left.\begin{array}{l} 0.3 < K_A < 2.0 \\ T_A = 0 \end{array}\right\}$ motor/rectifier in discontinuous current mode

$0.5 < \psi^* < 1.0$ — scaled field excitation

$330\text{ms} < T_H < 700\text{ms}$ — electro mechanical time constant

$0 < B < 2$ — friction coefficient (alterable with eddy-current brake)

Reference values: current 16 A
speed 1900 rpm
motor voltage 130 V
torque 10.4 Nm

Fig. 3.2 Scaled plant model (the signals within this nonlinear block diagram are considered to be in time domain, variable parameters: T_t, K_A, T_A, T_H and B).

Table 3.1 Technical data of the system

DC-Motor	
degree of protection	IP 23
rated power	1.1 kW
rated voltage	130 V
rated current	11 A
rated speed	1500 rpm
Tacho-generator	
degree of protection	IP 44
voltage	40 V/1000 rpm
current	10 mA, 40 mA max.
speed	1000 rpm, 9000 rpm max.
Eddy-current brake with phase angle control for the excitation	
supply voltage	220 V; 1~
frequency	50 Hz
rated output	1.0/1.3 kW DB/cont. duty
rated speed	1500/3000 rpm
Magnetic clutch and moment of inertia	
clutch max. torque	30 Nm
clutch max. speed	3400 rpm
motor + brake inertia	0.017 kgm^2
max. moment of inertia	0.036 kgm^2
AC-DC dual-converter for the motor armature	
supply voltage	220 V; 1 ~
frequency	50 Hz
rated DC voltage	150 V
rated DC current	15 A
rated power	2.2 kW
thyristors	SKKT15/08H1
AC-DC semi-converter for the motor field circuit	
supply voltage	220 V; 1 ~
frequency	50 Hz
rated DC voltage	180 V
rated DC current	15 A
rated power	2.7 kW

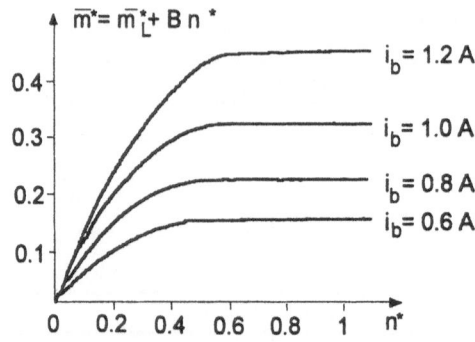

Fig. 3.3 Dependence of the scaled load torque ($\overline{m}_L^*$) and of the friction coefficient (B) with the excitation current (i_b) of the eddy-current brake upon the scaled speed (n^*)

3.4 References

Bystron, K. (1979), *'Leistungselektronik'*, Vol. II, Hanser Verlag, München/Wien.

Groenenboom, M. (1972), 'The evolution of firing circuits for static converters', Holectechniek, 2, pp. 1-13.

Harland, J. (1985), 'Aufbau und Erprobung einer Anpassung zwischen dem Drehzahlversuchsstand und einem Mikrorechner', Study thesis ESR-8507, Ruhr-Universität Bochum, Germany.

Langhoff, J. and E. Raatz (1976), *'Geregelte Gleichstromantriebe'*, AEG-Telefunken.

SIEMENS (1977), 'Demonstration Model for Variable-Speed Drive'. No. E 484/1517-220.

Wieser, H. (1985), 'Erweiterung der Anpassung zwischen dem Drehzahlversuchs-stand und einem Mikrorechner', Int. Report ESR-8507, Ruhr-Universität Bochum, Germany.

Wieser, H. and R. M. Stephan (1984), 'Das Demonstrations-Modell für Drehzahlregelung', Int. Report ESR-8319, Ruhr-Universität Bochum, Germany.

CHAPTER 4
THE MICROCOMPUTERS: HARDWARE AND SOFTWARE

4.1 Introduction

The implementation of a more complex digital controller is normally made with the assistance of a development system or a real-time operating system. Typically, an interactive dialog program is responsible for the establishment of controller parameters and structures, and it is necessary in order to test different controllers and to evaluate their performance and behaviour. The digital controller itself is programmed in the form of a real-time program. Algebraic and logic operations necessary for the control algorithm, as well as measurement of plant signals and the computation of control signals are the principal task of the real-time program. This program is usually scheduled at constant time intervals (the sampling time) and has high priority. The dialog program, which has low priority, is executed only at times, when the real time program does not use the CPU. These two programs share some variables, that make the dialog effective.

The realization of this scheme is often done with a single CPU under the control of a real-time operating system, that administers the usage of the CPU and its peripherals. The 16-bit microcomputer system available for the purpose of this investigation is based on a standard PC, which is not well suited for operation in an industrial control environment. Furthermore, the use of a general purpose real-time operating system for the control of the DC-motor has two significant disadvantages:

- The small sampling time in comparison with the computation time, that is necessary for the proposed control schemes, does not allow the CPU to be involved with operating system tasks.

- The interval timer, the interrupt controller, the parallel interface and the other peripherals necessary for the motor control have, normally under a general purpose operating system, already established tasks.

Therefore, the first step towards the realization of the controllers proposed here was to add a single-board-computer (SBC) computational system (Fig. 4.1). The first 8086/8087 SBC used in this project was based on the MULTIBUS I and was a product of the firm Matrox furnished with Intel components (Matrox, 1984). Also in the following prototypes the hardware was based on similar components but a simpler and cheaper bus system technology was used, which will be described in the next section together with the development system.

4.2 The Development System and the Single Board Computer

4.2.1 The Basic Concept

As stated in the introduction of this chapter, the use of a standard PC-based real-time system was not taken into consideration for the semi-industrial implementation of adaptive control for thyristor-fed DC-motors. Instead, a solution based on a standalone microprocessor system specialized for this control task was chosen. This meant also that in the beginning of the project no software was available to support the implementation. A development system was necessary to do the actual programming, including editing, compiling and linking of the control software. Existing development systems were found to be expensive or not appropriate for the programming in a mix of high and low level languages. The approach actually carried out in this project was the development of software with standard tools of the MSDOS/PCDOS operating system. These tools comprise the standard assembler, compiler and linker programs available for this system.

The complete system for development and application of adaptive control is depicted graphically in Fig. 4.1. A logic analyser can be connected to a maintenance port of the controller to monitor the status of the hardware and software. This is of importance for measuring the computation time of several tasks without interfering with the internal operation of the microcomputer, which would be the case if this would be done exclusively by SBC operating system software. A PC is connected by a serial communication line, which allows interactive supervision and operation of the control software by an intelligent terminal. The additional parallel connection can be used for fast download and upload of data in the development of control programs, or for the task of data recording and documentation during the normal operation of the control equipment. Documentation and terminal mode is supported by a printer connected to the PC for protocol purposes and hard copy of graphic screens. Analog and digital I/O lines form the interface to the controlled plant for controlled and manipulated values.

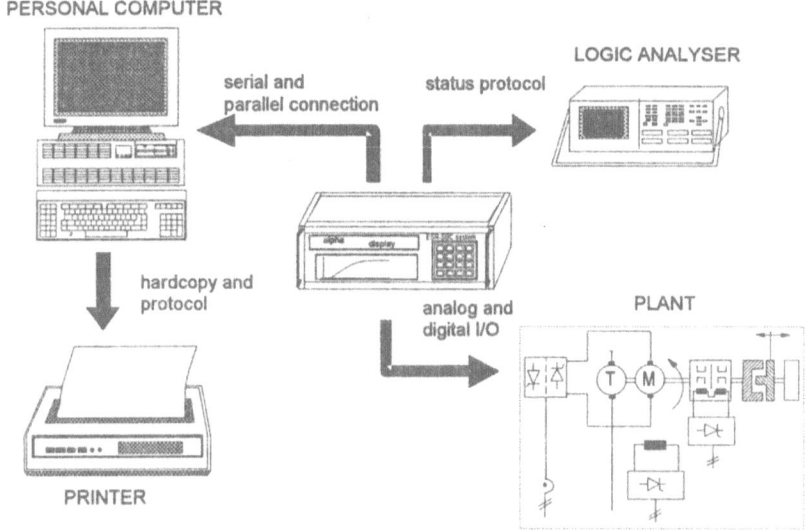

Fig. 4.1 The computer system for digital control applications

The hardware and the software necessary for the realization of the basic concept will be discussed in the next sections in more detail.

4.2.2 The Single Board Computer (SBC)

The first prototype of the SBC was based on the MULTIBUS I concept (Matrox, 1984). The analysis of the first implementation showed that the MULTIBUS has no advantages over other bus systems, when used in a standalone operation. The design of additional hardware for control purposes is more complex for the MULTIBUS than for 8-bit bus systems, and commercial components are more expensive. A 16-bit bus technology is actually not needed for an SBC, because no fast data paths for hard disk I/O or local area networking have to be provided. There is the necessity for a 16-bit memory data bus to access memory and coprocessor in a fast and convenient way. The standard peripheral support chips as timer (8253), interrupt controller (8259), serial (8251) and parallel interface (8255) of the 8086 microprocessor system are 8-bit components, well known from the 8085 8-bit microcomputer technology (Intel, 1978). Therefore, the decision for the extended Euro Card Bus (ECB), which had been in industrial use for the development of logical controllers in its non-extended 8085/Z80 version since the late seventies, was made. The extended ECB technology offers advantages as a simpler bus timing, interrupt and DMA handling, standard euro-card size for layout and routing in the design of interface cards, standard industrial 19" case, wide spread usage and support by special hardware tools during development and test. The experiences of public domain implementations as reported by Werner (1985) and the available knowledge of students and technicians with respect to this technology could be used. In addition, already available layouts for the interfacing of D/A-A/D converters and power supplies could be used with minor changes. The extension of the EC-bus consists mainly in adding 32 new bus signals (4 more address lines, 8 additional data lines, 8 interrupts request signals etc.) to the 64 standard signals, without interfering with the conventional ECB specification, so that available standard ECB 8-bit cards can be used as well.

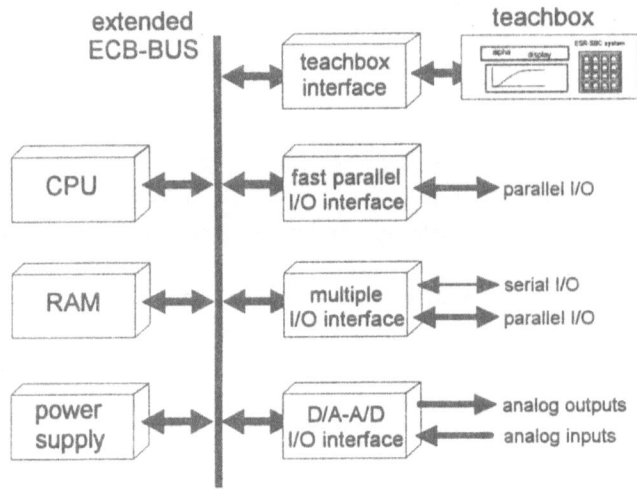

Fig. 4.2 The components of the SBC

The digital control equipment was first build around an 8086 microprocessor with clock frequency of 5 MHz. The volatile memory comprised 128 kByte of dynamic RAM. The connection to external devices is done by serial and parallel lines and by analog/digital and digital/analog converters, as illustrated in Fig. 4.2. The digital controller is equipped with a teachbox consisting of a small numeric keyboard with some additional function keys, an alphanumeric liquid crystal display and a graphical pixel-oriented LCD. The operating voltage is provided by a switching power supply with voltages ±5 V and ±12 V. The devices are grouped on separate removable cards equipped with the standard 96-pin ECB connector.

The centre of the microcomputer is formed by the *CPU-board*. In the second stage of the project the Intel 8086 processor had been replaced by a NEC V30 (NEC, 1988), which has several advantages. The power consumption of this CMOS processor is lower and it is nearly software compatible to the 80186 instruction set. The CPU and its peripherals could be driven by an 8 MHz system clock. Due to the dual processor mode - microprocessor and arithmetic coprocessor are sharing the same bus signals - it is necessary to operate the system in "maximum-mode" (Rector and Alexi, 1982). This means that a 8288 bus controller must be used. To be able to handle several interrupts, the CPU-Board was equipped with an interrupt controller of the type 8259A, which allows beneath other features, as prioritized interrupts, the multiplexing of the single CPU INT input for 8 different channels. For the bootstrap mechanism and the basic input-output software as well as for the user supplied control software two EPROMS offering a memory capacity of 64 kByte are provided.

Table 4.1 PC-MATLAB benchmark times (small values: good performance)

Microcomputer	multiplication	inversion	eigenvalue	FFT	LINPACK
(8MHz 68000)	76.0	101.0	107.0	91.0	256.0
(6.0MHz/80286/87)	12.1	19.5	17.5	13.7	45.8
(4.7MHz/8088/87)	9.5	20.9	21.0	12.3	40.9
(8MHz/8086/87)	5.2	11.6	11.2	7.1	23.5

The question why not to upgrade to the 80286 processor technology has a simple answer. For adaptive control problems the computational speed is of importance and the requirement of memory address space will not grow above the 1MByte boundary. Regarding the speed of computation, the 8086/8087 is superior to an 80286/80287 system, which can be concluded from the benchmark results in Table 4.1 (Mathworks, 1989), where even a standard 8088/8087 combination operated at a clock speed of 4.77 MHz is faster than a 80286/80287 processor/coprocessor system working with a 6 Mhz clock. This is especially due to the fact, that the coprocessor uses only 2/3 of the clock speed resulting in a 4.3 Mhz 80287 clock rate. Additionally, it is not possible to run 80286 and 80287 in parallel, because the arithmetic coprocessor is not able to supply the necessary bus signal to do addressing and data transfer by itself.

For complex control algorithms a sufficiently large amount of read/write memory is necessary from the developers and the industrial point of view. Not only for storage of data, but also for the running program code, RAM memory will be used. The access time of standard dynamic RAM chips is in the range of 100ns, whereas the access time for ROMS and EPROMS is three times as high, so that the software will be slowed down extensively if the code is fetched from ROM addresses. To meet this requests the system is equipped with a *RAM board* which allows a capacity of up to one MByte, in steps of 128, 256, 640 and 1024 kBytes and depending on the type of dynamic RAM chips used. The chosen configuration uses 16×64 kBit DRAM in bank 1, so that the total amount of random access memory corresponds to 128 kByte. This size is necessary, because it allows to load the contents of the 64 kByte EPROM memory to RAM and leaves additional 64 kByte free for data storage. This meets the so called "small memory" model defined by Microsoft (1985), where a segment of 64 kByte is used for program code and further 64 kByte are shared by data, stack and heap storage. The addresses of the installed RAM are in the range 0 to 1FFFFH. A hidden refresh implemented by hardware removes memory refresh duty cycles as they are known from the standard PC, where one counter and one DMA channel are utilized to implement a programmed refresh which consumes between 5 and 10% of the available processor time.

The microcomputer for digital control shall not only work in stand-alone operation, it should also provide the possibility to connect a terminal, allowing thus a comfortable interactive dialog with the user. For that purpose, a serial connection according to the RS232 standard is necessary. To display the status of the microcomputer, some LEDs driven by a parallel port should be available. In addition, the parallel interface can be used for free programmable digital inputs and outputs, e.g. in this project they are necessary to generate the thyristor firing pulses and to implement the input for discontinuous current detection. To generate a time base, produce cyclic interrupts and to provide the clock frequency for the serial communication line timer/counters are needed. Also the generation of the firing pulse, which requests an accuracy in the sub-millisecond range, should be produced by a hardware timer. Utilizing the space offered by an EC-board in an optimal way, a *multi-I/O-board* comprising two serial ports (8251), one parallel port (8255A) and two programmable timers (8253) has been constructed.

For DC-motor control a maximum number of two A/D converters, one for the measurement of the speed value offered by the tachogenerator and the other for measuring the current mean value, are necessary. The manipulated value in hybrid cascade speed control, namely the setpoint for the armature current, is the only analog value to be generated by the digital controller. So an *A/D-D/A converter I/O board* has been designed, which offers twice the number of interface lines as necessary. The 12bit analog to digital converter, based on the A/D converter AD 574 AK, sample&hold circuit AD 585 and the analog multiplexer AD 7507, offers

4 bipolar inputs for measurement. The conversion time of this circuitry is less than 25 μs. The digital to analog converter employs the D/A converter AD 667 AK, which is a single-chip solution taking 4 μs for the conversion of a 12 bit value (Microsystems Components Databook, 1985).

The SBC is also equipped with several TTL-based digital input/output ports, which serve for interfacing to the teachbox and to provide the fast parallel connection to the supervising PC.

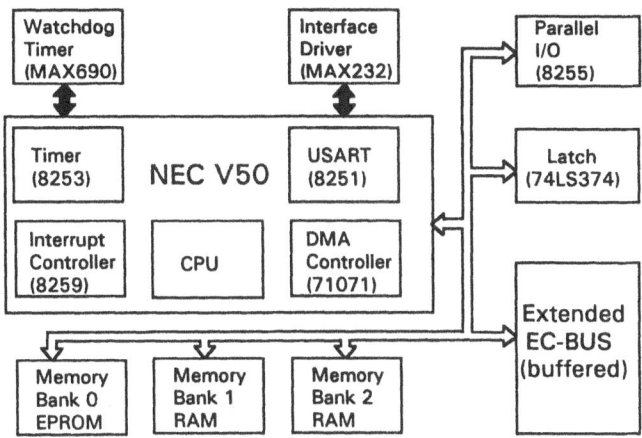

Fig. 4.3 The V50 based SBC concept

In the latest implementation, the basic equipment comprising CPU-board, RAM-board and I/O-board described before, has been replaced by a single printed circuit card based on the NEC V50 Processor (NEC, 1988), depicted in Fig. 4.3. This processor offers a 8253 compatible timer, the interrupt controller and the serial interface on chip, so that the I/O-board has become obsolete. By using static RAM technology, which does not need any complex refresh circuits, it was possible to place also 128 kByte RAM on the CPU-board, so that in addition to this board only one card containing A/D-D/A-converters and parallel interfaces to drive the teachbox and the parallel connection PC-SBC is needed.

4.2.3 The Teachbox

For operation without terminal the control system needs a user interface, which allows the input and the display of data. For applications in the field of control engineering the display of several measured or internal controller values is of great importance. The measured values can be quite different from case to case: If a digital current controller is implemented a display of its manipulated variable and the current should be possible. This is not necessary, if the current controller is implemented using analog techniques. This requests a more universal concept for the display. A graphical display based on a pixel oriented LCD is well suited for this task. Only the size, resolution and the speed of its hardware sets limits to its usage. For a simple command interface an alphanumeric display and a keyboard with numeric keypad and several function keys are sufficient.

This user interface has been realized by a teachbox comprising two liquid crystal displays and a small keyboard with sixteen keys. It is connected by a 70cm 40-conductor flat wire, realizing the parallel connection with the teachbox adapter card in the microcomputer. For normal operation the teachbox is attached to the 19" case of the microcomputer and can be detached from here for longer editing sessions.

The display incorporates a two row alphanumeric display with 40 characters per line and a graphical LCD with a resolution of 240×64 dots of type Hitachi LM018L and LM200. A schematic diagram of the teachbox is shown in Fig. 4.4. Both displays contain build-in controllers, making their programming quite comfortable and simple, so that a standard BIOS interface for both types of interfaces could be included in the machine-language-coded monitor program of the microcomputer.

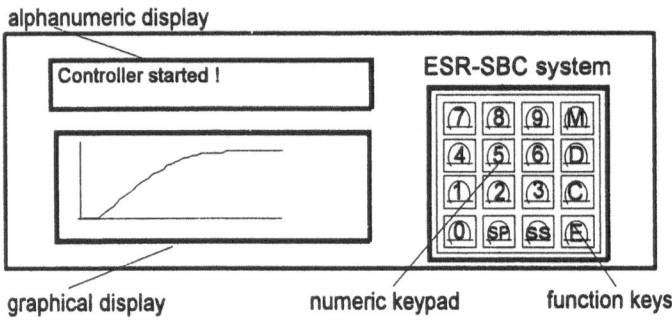

Fig. 4.4 Schematic diagram of the teachbox's displays and keyboard

The keyboard with 16 keys contains a combinatorial logic which decodes the key presses completely and sends 4-bit character codes. A strobe signal produced by each key press is used to initiate an interrupt service routine, which handles the keyboard input. Table 4.2 resumes the standard function keys. The adaptive control software should be written in such a way, that beneath start, stop and specification of the set point only one knob is necessary for the specification of the dynamics of the control loop. Depending on the actual control algorithm in use, the function key 'D' could for example be used to ask for the settling time, the damping ratio and characteristic frequency or for a weighting factor to specify the closed loop dynamics.

Table 4.2 Standard definitions for the function keys

key code	function
M	Used to invoke the *menu*, depending on the special control algorithm used. Hereby the control algorithms operational and structural data, e.g. sampling time, plant dead time and order, can be specified. Not used in normal operation.
D	Specify the dynamics of the closed loop system.
C	Cancel input.
E	Enter key used for numeric input etc.
SS	Toggle for start and stop of the control algorithm.
SP	Specification of the set point.

For the initialization and operation of the various devices described in the section on hardware, special software is necessary. Especially, if the hardware is under steady development, this basic software should be written and organized in such a way, that the problem oriented control software is completely detached from the hardware and device dependent addresses. This aim can be reached by a monitor program which forms the interface to the hardware in a way similar to the PC ROM-BIOS. The software of the basic input-output system will be described in the next section.

4.3 The Monitor Program

4.3.1 The Single Board Computer ROM-BIOS

The monitor program may be seen as the operating system of the SBC. It is a small program compared to a full-scale operating system, but it performs all the duties necessary to execute the control task and to support the development of control algorithms. During the phase of planning this minimal operating system, one tried to carry over the advantages of other basic software systems for the 8086 architecture of microprocessors, and to avoid their disadvantages with respect to real-time operations. Another goal was to keep the amount of memory used by the monitor small, thus leaving enough memory for the user program and for the purposes of digital control. The monitor program consists mainly of 3 blocks:

- Initialization of the processor, peripheral devices and memory,

- command level with service subroutines for upload, download and start of a user program,

- user's library for analog/digital input-output, communication and control of the real-time program.

The first version of this program had the only task of bootstrapping the system and preparing for a download of more efficient service routines or any control program, which incorporated also the I/O library. But in this case the problem oriented user program would become machine and device dependent. If only minor changes in hardware I/O addressing are made, the digital controller has to be partially recompiled and relinked. The monitor program offers a method for device and hardware independent programming, so that a digital controller once compiled will run without any changes on all hardware platforms as there are the earliest MULTIBUS system, the ECB system with several boards and the V50 solution with one board for processor, RAM and I/O.

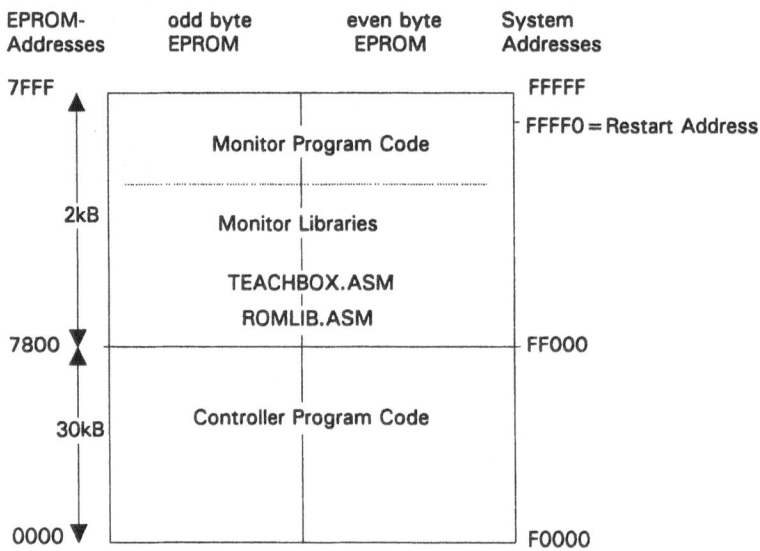

Fig. 4.5 Memory map of the monitor EPROM

If the voltage supply of the SBC is switched on, a reset pulse is launched for the microprocessor. The program counter will be set to 0FFFFH where the monitor program starts. This address must contain a jump instruction into the initialization code. To avoid any spurious interrupts owing to noise on the peripheral data lines, the interrupt flag of the processor will be cleared and the initialization procedure is started. After testing all processor registers, the segment registers will be set to

42

standard values for the test on existing memory. If this test is passed, all peripheral devices are checked and initialized. The numeric coprocessor will be reset and its control word is set.

The interrupt controller will be programmed for the correct mode and the mapping of the interrupt vectors will be carried out after the pointers for these vectors in zero page memory are set to valid interrupt service routines in the monitor code segment. In the sequence interval timer, serial interface, parallel interface, D/A-A/D-converter and teachbox interface all peripheral registers are set to default values, so that the I/O library can handle this devices. After the successful initialization of the peripheral interfaces, the software environment for the monitor and the user's program has to be prepared. Up to now the monitor executed his code residing in the EPROM, as illustrated in Fig. 4.5. Only a small amount of RAM for stack memory is used during testing and initialization, which can be monitored on the LED display and on the diagnosis port. If an error occurred, a specific pattern on the LED will signal the reason for the failure.

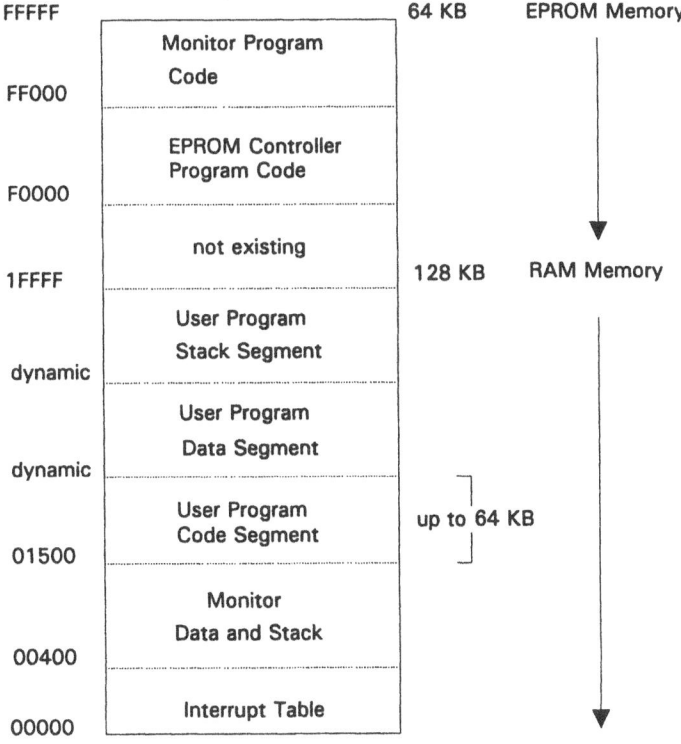

Fig. 4.6 Memory map of the SBC

The two 256 kBit EPROMS contain not only the 4 kByte monitor program, but also 60 kByte of controller program code. It is the task of the monitor to move the controller program code to RAM and to start this program. The memory map of the microcomputer with loaded program for digital control is illustrated in Fig. 4.6. The monitor uses the interrupt table and a small amount of RAM for data and its stack area at the bottom of the installed RAM. At the hexadecimal address 1500H starts the code segment of the user program providing up to 64 kByte of memory, which is followed by the data segment, the stack starts at the end of the installed RAM and grows downward.

If a power-on reset cycle is executed, the user program is started immediately after initialization and relocation. But during development and test, the loading of a program directly from the PC and not from the built-in EPROM is the standard case. This mode of operation can be reached by pressing the reset button. During the reset cycle the monitor checks whether it is a cold reset. This is simply done as in the PC BIOS, a special word in the RAM is checked for the value 1234H. If this value exists, a warm boot is executed and the monitor enters a communication loop polling the parallel connection to the PC and waiting for the specific commands packets LOAD, SEND and GO. The LOAD command can be used to download a program from the PC to the SBC, the SEND command serves for uploading a copy of the SBC memory to the PC and GO starts the program operation at a specified address.

Table 4.3 Service routines of the ROM BIOS

group	service	software interrupt
1	D/A and A/D conversion	60H
2	parallel interface	61H
3	serial interface	62H
4	timer	63H
5	operating system	64H
6	fast parallel interface	65H
7	teachbox	66H

The ROM BIOS of the SBC also offers routines to control the interface devices. According to Table 4.3 the services are divided into 7 groups according to the special hardware devices. Each group comprises several requests, e.g. initializing, reading and writing from and to the specific devices. The operating service offers routines to handle hardware interrupts and to control the scheduling of the real-time tasks.

Calling a monitor service is very similar to calling a DOS or BIOS routine on a PC. According to the request, the high byte of the AX register is loaded with the request code other registers of the processor are loaded with data or pointers to the data. Finally, the software interrupt is executed. Writing a string to the alphanumeric display could be done for example as indicated in the Pascal code of Table 4.4.

Table 4.4 Sample Pascal code for using the ROM BIOS

```
Procedure Hello;
Const
  HelloStr = 'Hello world!';
Var
  Regs : Registers;
Begin
  Regs.AX:=$0200;          { Request code alpha display  }
  Regs.CX:=Ofs(HelloStr);  { Offset and segment of the   }
  Regs.DX:=Seg(HelloStr);  { display string to DX:CX     }
  Intr($67,Regs);          { Exec teachbox soft interrupt`}
End;
```

It is simple to access the BIOS service routines from any language supported by the SBC. For the Pascal and C programming languages there exist predefined library functions, which are part of the general support libraries, that will be described in section 4.5.

4.3.2 The Real-Time Scheduler

The fast sampling time, necessary for the control of electro-mechanical plants, does not allow the usage of any of the more sophisticated multi-tasking mechanisms. In our system only one task is allowed to be scheduled as controller in the intervals specified by the sampling time. All other tasks are bound to hardware interrupts and should queue their results by themselves, thus making the information available for the controller task and the main program code, which is performing the idle loop. The tasks bound to hardware interrupts, as for example the PI current controller or the interrupt service routine for the teachbox and the serial/parallel interface, finish their operation in a nearly negligible time compared with the resolution of the scheduler, which is just 1 ms. And they all do not access the numeric coprocessor, so that the response time with respect to a hardware interrupt can be kept in the range of 50 to 100 microseconds.

So it is possible to work with a very simple scheduler routine for the management of the adaptive speed controller. The scheduler is attached to hardware interrupt #3 which is always executed, if a cyclic counter in the 8253 interval timer has expired. The counter register is initialized in such a way, that it needs 1 ms to count down to zero. Multiples of this time base can be used as sampling time. It is not possible that the hardware counter can be used directly to schedule the control algorithm, because it can happen that the control algorithm is delayed by other tasks running in the system, so that it is still active, when the new scheduling takes place. If a hardware timer would be used, the scheduling for the next sampling step would actually take place after a dead time equal to the sampling time. In case of a scheduling mechanism which uses an additional software counter, this dead time is equal to the granularity of the counter of 1 ms, so that large delays can be avoided.

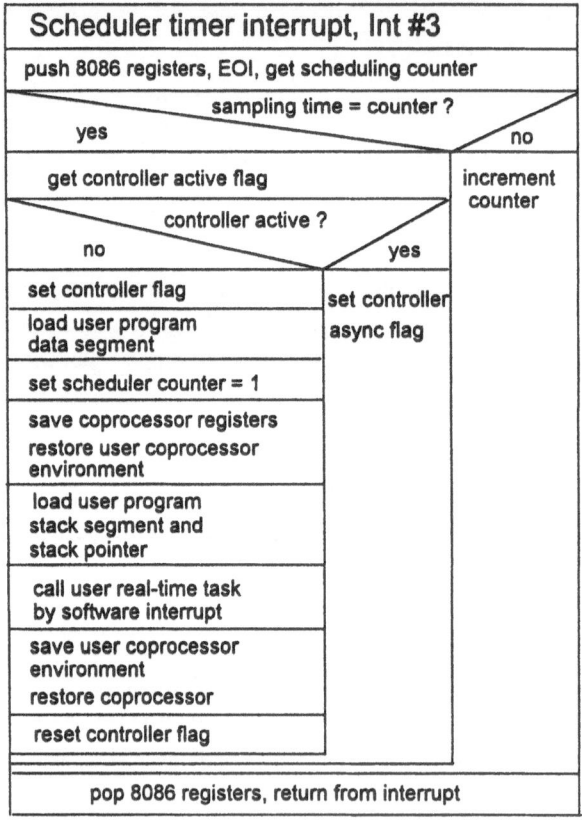

Fig. 4.7 Program flow in the real-time scheduler

Fig. 4.7 presents a simplified flow diagram of the scheduling mechanism. When the hardware interrupt is executed, the 8086 registers are pushed and the "end of interrupt" command (EOI) is sent to the interrupt control to be able to receive further interrupts. Then the software counter is checked, whether it has expired. If it is not the case, it will be incremented and the scheduling routine returns from interrupt processing.

If the counter has reached the sampling time of the control algorithm, the environment of the user program, i.e. stack and data segment, stack pointer and coprocessor status, will be restored to the state of operation requested on the installation of the real-time task.

4.4 The Interface to the CADACS System

4.4.1 Overview to the CADACS System

Computer-aided analysis and synthesis uses the computer as first-rank tool for the solution of a given problem. For the handling of practical problems a broad spectrum of methods and tools has to be offered, which benefits the developer of algorithms and methods as well as the later user. The scope covered by these tools is modelling, analysis, controller design and simulation. These tools have been utilized, when preparing the implementation of the adaptive control algorithms for DC-motor control. Furthermore, they supported the experiments and their evaluation.

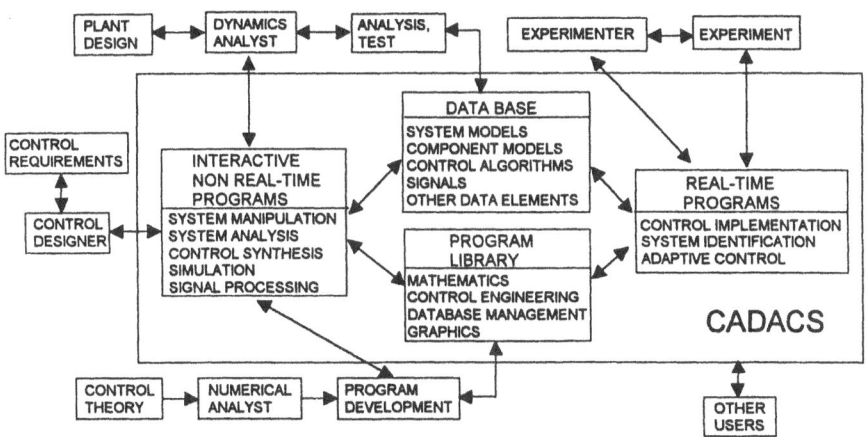

Fig. 4.8 CADACS applications environment

CADACS (Schmid, 1985; Unbehauen, 1987; Unbehauen et al., 1990; Keuchel and Schmid, 1991) has been designed for a broad area of applications in the field of control engineering problems. The modular program system comprises besides classical and modern methods for analysis and synthesis of SISO and MIMO control systems, simulation as well as implementation of control algorithms in real-time. A comprehensive library of methods, the flexible facilities for graphical display and state-of-the-art numerical algorithms are characteristic properties of a CADACS working environment. The application system is tailored for the user, but it can be extended using the library or source program option by user defined modules. The real-time toolbox offers, in extension to the test operation of controllers designed using the system, the development of new controller software.

The block diagram, Fig 4.8, illustrates the main components of CADACS and their relation to the application area. The most important components consist of a group of *interactive programs* for the management of dynamic systems, for system analysis, controller synthesis, simulation and signal processing. A unified *data base* for system models allows a smooth exchange of data between the different programs and in between different groups of a development or research project.

4.4.2 The Data Base and Types of System Representation

Often needed subtasks for analysis and synthesis are supported in a centralized way. This is valid for example for the managing of numerical system representation forms and the transformations from one form to another. In CADACS, ten different types of representing a system in frequency and time domain are available. Identification, simulation and also approximation are considered as transformation to a different system representation (Schmid, 1985).

A large number of alternative ways for modelling, computation, transformation and simulation offer a high degree of flexibility, which is necessary to handle the broad spectrum of tasks occurring during control engineering projects. For example the result of the modelling phase can be transformed to a representation which is better suited for controller design.

According to the different types of system representation and components of dynamic systems, the data base offers for each type of control engineering object a specific binary type of file. In addition further supplementary objects are necessary. These exist for: signals, transfer functions (matrices), coefficient matrices, state space systems, frequency tables, frequency response tables, spectra, descriptions of the graphical layout, structure descriptions, control engineering objects in textual format. In the present project signals, transfer functions and descriptions for the graphical layout have been used extensively.

4.4.3 Tools for Kernel Functions

Standard tasks of moderate size are grouped together to modules associated with the type of system representation. The necessary tools are realized by central interactive programs, so called Managers (see Table 4.5). Using these Managers, transfer systems may be defined and handled easily. Two of these managers are of great importance with respect to the project of real-time control based on SBC. One is the Graphics Manager which served not only for the graphical display of signals, but also offered the conception for the interface between PC and SBC. In an early stage of the development of CADACS, when implementation of this package on different hardware and operating system platforms was carried out, one decided to divide interactive programs and graphical display by definition of a mailbox oriented interface.

Table 4.5 The managers of CADACS

Signal Manager SDISP	Handling of signals and computation of signal characteristics.
System Manager SMGR	Handling and analysis of transfer functions in s- and z-domain.
Frequency Manager FRMG	Handling and analysis of frequency response tables, entries of frequency response matrices or spectra.
Matrix Manager MMGR	Handling and analysis of systems in state space representation.
Graphics Manager GRMGR	Centralized graphical representation of control engineering objects in diagrams.
Documentation Manager DOKMG	Documentation of signal sequences generated by real-time experiments.
Cross Manager XMGR	Import and export to and from the standard binary data base.

All tasks for graphical representation within CADACS are handled by the Graphics Manager. This module can be seen as a graphical device driver, which is able to transform the graphical information generated by CADACS programs into diagrams. The principle of operation of the graphical display is depicted in Fig. 4.9. The information to be represented graphically, including a diagram request, is put into a mailbox. Graphics Manager takes this information and maps it to a specific diagram.

The user defines by a so called 'problem menu' the form and type of graphical representation for his problem-oriented data. For standard representations a menu library is available, which can be extended for each individual user or for a specific application. For that purpose a diagram definition language is available. This form of centralized organisation of graphics display allows different programs to map clearly and in a unified way into a single diagram or to map results on a single graphics working sheet in several different diagrams.

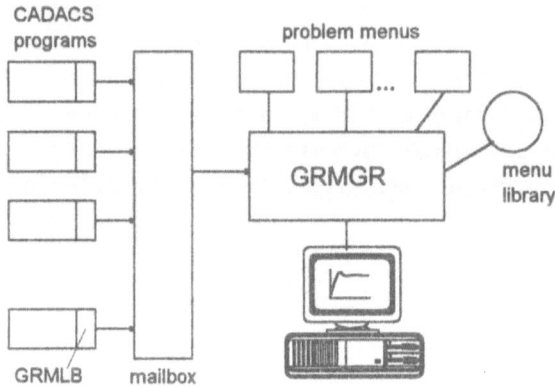

Fig. 4.9 Function principle of the Graphics Manager

The mailbox, which uses a specific protocol and which can be implemented in different ways, e.g. by synchronous or asynchronous mailbox I/O provided by the operating system, by system FIFO buffers or by piping structures, makes the application programs independent of the specific graphics hardware and software. The graphics display system could even be implemented on an intelligent graphics terminal or on a different workstation.

On a PC using the PC/MSDOS operating system, the mailbox has been implemented by software interrupts similar to the mechanism of DOS requests. The enhancement of this communication and programming mechanism with respect to hardware interrupt routines, where the interrupts are generated by a remote computer was very simple and did only request minor changes in the available software.

Also the Documentation Manager was implemented in a similar way interactive CADACS programs assemble a message buffer, which is actually a C struct or a Pascal record data structure and allows simple access to the contents of the message. With the AX register set to a number defining the request and with the register pair DX:CX pointing to the message buffer a software interrupt is called. The handler for this interrupt is part of the Documentation Manager which is a normal Pascal procedure of type interrupt. The interrupt handler gets the request code from the AX register and moves the message pointed to by DX:CX to a local buffer and processes it. When this processing is finished or when the message is queued, it returns from the interrupt, signalling the success or failure by the contents of the AX register. This mechanism internal to the PC-implementation of CADACS can be generalized with minor changes for the remote operation of real-time control software. Instead of using software interrupts, the communication is executed by hardware interrupts. In the case of data documentation via the fast parallel connection, hardware interrupt #7 was used. And instead of passing the request code in a register, it has to be part of the message which is transferred by a hardware handshake protocol using strobe and acknowledge signals. An additional software protocol based on evaluation of 8-bit checksums signals the success of the communication operation.

4.4.4 Tools for Analysis and Synthesis

For more complex tasks in analysis and synthesis specialized tools in form of closed programs are provided. Complete techniques are therefore compactly grouped together in a single module, which allows, initiated by the user, to carry out the computer-aided analysis or synthesis in whole.

Table 4.6 CADACS programs for system identification

PTAPP	Approximation of step responses or impulse responses by PT_n transfer functions with prespecified order or automatic search of order.
PRONY	Approximation of step responses or impulse responses by exponential functions with order detection. Laplace transformation and computation of transfer function.
MLID	Parameter estimation for SISO and MISO systems using the least-squares or maximum-likelihood method.
IMIMO	Parameter estimation for MIMO systems described by left coprime polynomial matrix fractions using the least-squares method (pseudo recursive and non-recursive).

These programs often offer special facilities for simulation, necessary for verification and judgement of the results. This does not exclude that, due to any necessity for extended simulation, the standard simulation tools can be used in the course of analysis and synthesis. The programs for identification and controller design are resumed in Tables 4.6 and 4.7.

Table 4.7 CADACS programs for controller design

OBSDN	Design of observers, full- and reduced-order, continuous- and discrete-time, with and without disturbance feed-forward, disturbance observers, by a pole assignment approach.
RICAT	Design of linear quadratic optimal proportional or proportional-integral state feedback controllers. Design of stationary Kalman-filters.
ROPTI	Parameter optimization for general discrete-time controllers, PID controllers for tracking and regulation. PID controller design according to tuning rules.
WOKU	Root locus generator for computation and graphical representation of root locus plots for various parameters for continuous- and discrete-time systems.

4.4.5 Programs for Real-Time Operation

CADACS has its roots in the implementation on mini-computers like HP1000 and PDP11, which provide with RTE-IV and RT-11 real-time operating systems. After the first microcomputer implementation using IRMX as operating system, also a PC implementation under MSDOS was established (Unbehauen et al., 1990). As illustrated in Fig. 4.10, this implementation parallels the minicomputer versions. The real-time task is a separate program, the exchange of data between real-time task and the supervising interactive user program is done through a shared common memory area. The requests to the real-time monitor as well as the scheduling of the real-time task is based on the concept of software interrupts.

Table 4.8 CADACS programs for real-time operation

DATST/DATAC	Data acquisition with and without generation of test signals.
ZSTEW/SFEDB	State feedback controller with and without observer for multivariable systems.
DDCST/DDC	SISO control loops with and without setpoint feed forward.
APPCS/APPCA	(Decentralized) adaptive pole placement control or LQ(G) optimal control.
MRACS/MRACA	(Decentralized) adaptive model reference control.

The task of the operating system is in the MSDOS implementation performed by a small assembly program, the real-time monitor, which is responsible for administration of the CPU in this two-task environment. Besides the real-time monitor a pair of programs is necessary for the implementation of different types of controllers. The so called 'user program' serves for implementation of the control program, for manual changes, for documentation of results and for signal monitoring in combination with DOKMG and GRMGR.

How the message system implemented by software interrupts can be carried over to a messaging scheme using a physical communication line has been reported in section 4.4.2 for the Documentation Manager. The exchange of data by a shared common structure is no longer possible, if the real-time task is running on a remote microcomputer. But utilizing the special implementation of user interaction in CADACS real-time programs, it is easy to modify the data exchange mechanism for remote operation.

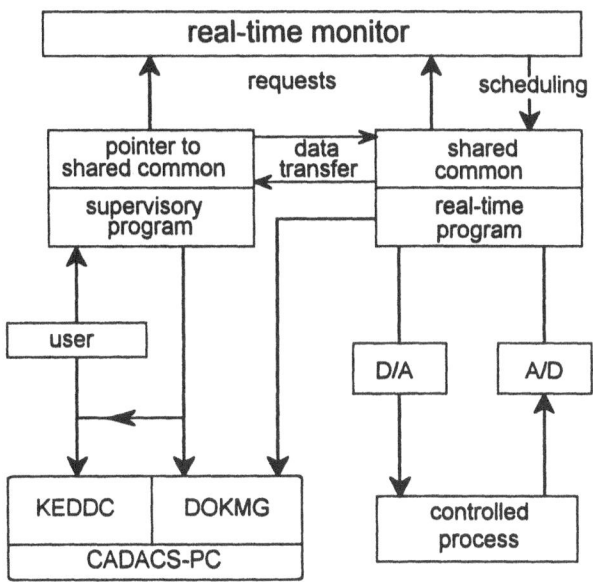

Fig. 4.10 Real-time programs in a process computer environment

The user input to real-time programs is always buffered in CADACS programs. The variables residing in the shared common are never referenced directly by user input. If the user enters a command, for example to change the setpoint, the attached value is entered into a buffer variable. After all interactive commands, specified by a chain of commands, have been executed, the buffered values are transferred in one chunk to the shared common variables, which are used by the real-time program. This serves to guarantee the consistency of the whole data set. Otherwise it would be possible to specify for example a new sampling time and a new set of controller parameters. If the altered sampling time would come into effect directly on input, the actual controller would be executed with the changed incorrect sampling time. This inconsistency could lead to instability of the control algorithm.

The concept of buffering the data of the interactive user input makes it easy to substitute the exchange of data via a shared common by the exchange of data using a messaging mechanism. Instead of moving the changed data to the shared memory partition, they are send via a protocol over the serial communication line. So the standard interactive programs have only to be changed with respect to the data exchange and can be used also for the supervision of remotely executing real-time tasks.

4.4.6 The User Interface

It is the task of the user interface to render possible the interactive usage of the package. It is necessary to offer a unified interface for all tools of a CAD system, to simplify the logical chaining of different steps, which are involved in the solution of a specific problem. Results will get evidence by graphical representation. The graphics interface should be almost independent of the hardware used. The user should be able to use predefined standard diagrams as well as diagrams designed by himself.

The handling of the system shall be clarified by the following commented example of dialog. During implementation of a real-time model reference controller the step response of the plant shall be examined graphically. For that purpose the simulation problem is interlaced into the dialog of controller implementation.

Prompt Command	Comment
MRAC: @@	main program call from real-time environment
KEDDC: DG	command 'digital simulation'
DIGSI: RE	local command for reference to data base
FILE NAME FOR PLANT = ? STR.UEB	specification of a file containing the plant
DIGSI: UF	local command to generate step response
SIMULATION TIME =? 10	specification of additional information
DIGSI: EX	end 'digital simulation'
MRAC: @@	dialog for implementation

CADACS allows a free and simple type of dialog. Using unified and very basic commands from the overall set of main tasks, one task, i.e. 'Digital Simulation', is chosen. Having done that, all commands of DIGSI are available to the user. With these commands, consisting of two characters, a subtask is scheduled from within DIGSI, which can be a calculation or a local question-and-answer dialog. A standard set of commands allows to call off status information, command menu, protocol or graphical information. In addition, a help facility is available, which offers to the user detailed analysis of errors with explanation and help. Each subtask is callable at any time in such a way, that a running main task is suspended and continued at a later time. The system is divided into several single programs. The exchange of data is made by files. The central program control and a number of utility commands is executed by the main module KEDDC.

Although the combined command and question-and-answer dialog seems to be complicated, it will be simplified by an intensive program guidance and support. Using *??* a local menu of commands will be displayed. In the case of an error the *HE*lp command displays an explanation of the last error with an additional error help and support to correct the situation. Furthermore, the user does not need a printed manual, because it is available on line, offering more than 1300 screen pages of information. During dialog each page of the manual can be called, giving comments to the methods used, and explaining the principle of operation of a specific program.

4.5 The Support Software

4.5.1 Generation of Programs with a PC

The data exchange between SBC and PC, the efficient application of the floating point coprocessor, and the comfortable use of the SBC peripherals are supported by several libraries created during this investigation. The so called ROMLIB, available in the high level programming languages C and Pascal as well in form of assembly language macros, serves as a convenient way to get access to the ROM BIOS routines discussed in section 4.3.1.

Over the last ten years several PC compilers for high level programming languages have been used for the development of control software for stand-alone SBC systems. Beneath Pascal compilers (Digital Research, 1983; Polydata, 1983) also the C programming language (Microsoft, 1987) has been applied for the implementation of controllers. Here only the approach for one specific Pascal compiler, the PolyPascal system (Polydata, 1983), that was actually used for implementation of the controllers presented in chapter 5, will be shortly discussed.

An executable program compiled and linked with standard PC software is not able to run on a different microcomputer system, which does not provide the services of MSDOS and the PC BIOS. For our implementation of control software no direct calls to DOS services will be made, but DOS plays an important role in loading and start-up of a program.

The executable produced by the MSDOS or compatible object file linker consists of two parts:

- the control and relocation information header,

- the load module itself, i.e. the user's program.

The control and relocation information are used by a loading program to set the initial values of the CPU segment registers and to do the necessary segment address relocations, before the user's program is started (Microsoft, 1991). The DOS operating system contains a loading program (the EXEC request, INT 21H, AX=4900H) to prepare .EXE files to run on a PC. To run .EXE files on the SBC, it was necessary to create a relocating loader, that relocates the load module before it is download to the SBC or burned into the EPROM. Because the segment start address of the user's program on the SBC is always fixed to 1500H, no relocation facilities have to be included in the monitor program of the SBC. Fig. 4.11 shows a diagram describing the generation of programs for stand-alone operation.

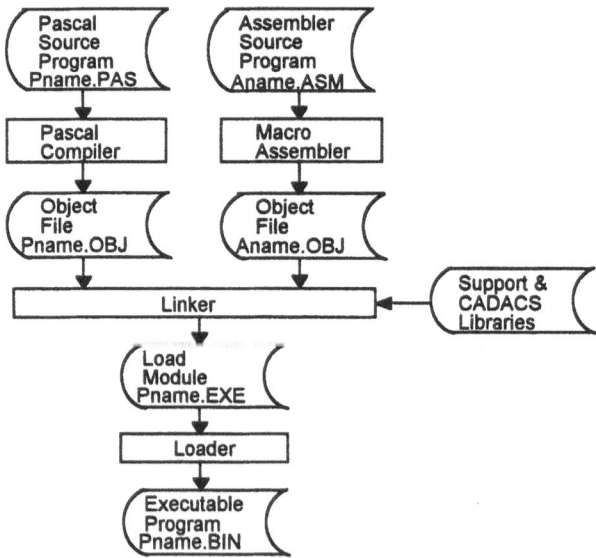

Fig. 4.11 Generation of an executable program for the SBC

The relocating loader does not solve all problems in preparing a DOS load module for execution under monitor control. The start-up module of the Pascal run-time library requests DOS to set interrupt vectors for emulation of the coprocessor, critical error handling and non-maskable interrupts. Additional requests for dynamic heap memory allocation and access to DOS environment variables are other typical actions during start-up. There are two possible approaches to solve this problem: The operating software can simulate the DOS request handling, so that correct values are returned to the specific request, or the start-up run time module of the compiler package has to be changed, so that no DOS requests will be executed. In the case of the Pascal compiler the second approach was carried out.

The support software allows the development of programs for the SBC on a standard PC in a fast and comfortable way. The download to the SBC by means of the parallel connection is as fast as writing a file to hard disk storage. The only missing development tool is a debugger, but testing of newly developed algorithms should be better done directly on a PC. As debugging aid the monitor provides an exception handler. If an execution error occurs, an error number and the position of the program counter is displayed on the alphanumeric display of the teachbox. This information can be used to find the source line of the program, where the error occurred.

4.5.2 The Real-Time Toolbox

It is tedious to program, debug and optimize a numerical routine written in assembly language. But also with highly optimizing compilers the same speed and performance of basic software for interrupt handling and basic signal and data processing is not attainable. To keep the size of the code small and reach a high speed of operation for the monitor program and the basic input output system, it was decided to implement this part using assembly language. Owing to the time consuming assembly coding, it is not advisable to code the high level control and estimation algorithms, which use matrix algebra and iterative numerical algorithms, in the same way. So one started first to use the Pascal programming language for that purpose. After a short time of evaluation of this approach and first arithmetic experiments it was clear that the aim of implementing a highly sophisticated adaptive controller using solely high level languages could not be reached.

This can be illustrated by a simple example. Suppose, that the following operation with double precision floating point variables should be done

$$RESULT = VAR1 + (VAR2*VAR3) + (VAR4-VAR5) * VAR6 . \qquad (4.1)$$

The computation time for different approaches on a 5 MHz 8086/87 system is given in Table 4.9. The saving in computation time can even be increased, when vector operations are executed. In this case, not only loading and storing of operands to and from the arithmetic coprocessor is executed, but also the indexing operations involving integer additions and multiplications have to be optimized.

Table 4.9 Computation time to solve Eq. (4.1)

computation time	method
60.06 ms	PASCAL program linked with 8087 emulator
0.88 ms	PASCAL program linked with 8087 arithmetic subroutine library
0.41	PASCAL program using 8087 inline code
0.31	8087 assembly language program

Table 4.10 The library for vector operations

```
Invocation:
  <subr> ([scalar,] v1,incr1 [v2,incr2, [v3,incr3,]] n)

subr         VIS-Routine Name.
scalar       Operand or result
v1, v2, v3   First array element as operand or result
i1, i2, i3   Offset for next element
n            Integer for number of elements to work on

VABS(v1,i1,v2,i2,n)              v2 ← |v1|
VADD(v1,i1,v2,i2,v3,i3,n)        v3 ← v1 + v2
VSUB(v1,i1,v2,i2,v3,i3,n)        v3 ← v1 - v2
VMPY(v1,i1,v2,i2,v3,i3,n)        v3 ← v1 * v2
VDIV(v1,i1,v2,i2,v3,i3,n)        v3 ← v1 / v2
VSAD(scalar,v1,i1,v2,i2,n)       v2 ← scalar + v1
VSSB(scalar,v1,i1,v2,i2,n)       v2 ← scalar - v1
VSMY(scalar,v1,i1,v2,i2,n)       v2 ← scalar * v1
VSDV(scalar,v1,i1,v2,i2,n)       v2 ← scalar / v1
VPIV(scalar,v1,i1,v2,i2,v3,i3,n)   v3 ← (scalar * v1) + v2
VDOT(scalar,v1,i1,v2,i2,n)     scalar ← sum[v1 * v2]
VMAB(scalar,v1,i1,n)           scalar ← index of max[|v1|]
VSUM(scalar,v1,i1,n)           scalar ← sum[v1]
VNRM(scalar,v1,i1,n)           scalar ← sum[|v1|]
VMAX(scalar,v1,i1,n)           scalar ← index of max[v1]
VMIN(scalar,v1,i1,n)           scalar ← index of min[v1]
VMIB(scalar,v1,i1,n)           scalar ← index of min[|v1|]
VMOV(v1,i1,v2,i2,n)              v2 ← v1
VSWP(v1,i1,v2,i2,n)             v1 ↔ v2
```

A special feature of the combination 8086/8087 can in this case be used, when programming in assembly language, namely the parallel operation of microprocessor and arithmetic coprocessor. While the 8087 is processing a numeric operation, the 8086 is free for the index operation to address the next operands in the vector processing loop.

Therefore, a vector operation library has been programmed to support operations as they occur for example when forming dot products between parameter vectors and signal vectors, e.g.

$$u = p^T m = [\, p_1 \; p_2 \; p_3 \; \cdots \; p_n \,] \, [\, m_1 \; m_2 \; m_3 \; \cdots \; m_n \,]^T, \qquad (4.2)$$

or for the elementary equivalence operations like pivoting, scaling and swapping rows and columns of matrices in the course of solving linear systems of equations. A short description of the various subroutines of this library for vector instructions (VIS) is given in Table 4.10.

To be able to do the indexing operation also for row and column vectors of matrices the parameter list contains, beneath the address of the first entries of the specific vector operands, the increment to access the next consecutive entry. In Pascal, where matrices are stored in terms of row vectors, the increment to access the operands row-wise would be equal to one. To access a column vector, the increment has to be equal to the number of dimensioned columns of the array. If, for example, a Pascal program declared and initialized a matrix of the form

```
A : Array[1..10,1..10] of Real
```

and the operation

$$A = A + I,$$

where I denotes the identity matrix, has to be executed, this can be coded using vector instructions as addition of a real scalar value to the diagonal by

```
VSAD(1.0E0,A,11,A,11,10) .
```

To discuss the properties of the vector instruction set in more detail only the dot product operation as representative example will be presented here. For that purpose one has to analyse the resulting machine code produced by high level language compilers with the low level program created by an experienced assembly language programmer.

Table 4.11 indicates a C program for the dot product. Table 4.12 shows the resulting assembly code generated by a widely used C compiler.

Table 4.11 Code for the dot product in the C programming language

```
void vdot(double *s,double v1[],int incr1,double v2[],int incr2,int
n)
// Calculate the dot product of 2 vectors s = <v1,v2>
{
    int j2, j1, i;
    long double tmp;
    j1 = j2 = 0;
    tmp =0.0e0;
    for (i = 1; i <= n; i++)
    {
        tmp += v1[j1] * v2[j2];
        j1 += incr1;
        j2 += incr2;
    }
    *s = tmp;
    return;
}
```

When analyzing it, the problems of compilers doing optimization for the 8086 processor architecture become obvious. If a human optimizer starts programming, he would use the available resources, i.e. registers and coprocessor stack, in such a way, that he would use the advantages and avoid the disadvantages of the special processor/coprocessor architecture.

The 8086 offers two general purpose pointers, formed by the register pairs ES:DI and DS:SI, which are very useful in string or vector operations and one up-down loop counter, the CX register. The compiler does not use this facilities as can be concluded from Table 4.12. The pointer to access the vector entries is always newly established using the ES:BX register pair, the loop variable occupies the misused SI index register and CX is not involved in any way. In addition, the resulting dot product is accumulated in memory and not in top of stack of the arithmetic coprocessor.

How an optimal code could be produced, shows the human coded assembler program in Table 4.13. Here, the correct use of the architecture of processor and coprocessor leads to a small-size and high-speed realization of the dot product. As loop count serves the CX register, ES:DI and DS:SI play their role as pointer to the vector entries, DX and BX contain the byte adjusted increment value and the accumulation is done in the top of stack of the coprocessor. The accumulated sum is stored to memory after the loop has been executed. While the coprocessor is adding the operands by the FADD instruction, the new index values can be established and the loop count can be tested in parallel operation of the 8086 microprocessor.

60

Table 4.12 Code for the dot product produced by the C compiler

```
_vdot     proc   far
    enter 10,0
    push  si
    push  di
    ;             j1 = j2 = 0;
    xor   ax,ax
    mov   dx,ax                  ;j1
    mov   di,ax                  ;j2
    ;             tmp =0.0e0;
    FLDZ
    FSTP  tbyte ptr [bp-10] ;tmp
    ;             for (i = 1; i <= n; i++)
    mov   si,1                   ;i
    FWAIT
    jmp   short @1@98
@1@50:
    ;             tmp += v1[j1] * v2[j2];
    mov   ax,di                  ;j2
    shl   ax,3
    les   bx,dword ptr [bp+10]
    add   bx,ax
    FLD   qword ptr es:[bx]
    mov   ax,dx                  ;j1
    shl   ax,3
    les   bx,dword ptr [bp+16]
    add   bx,ax
    FMUL  qword ptr es:[bx]
    FLD   tbyte ptr [bp-10] ; tmp
    FADD
    FSTP  tbyte ptr [bp-10] ; tmp
    ;             j1 += incr1;
    add   di,word ptr [bp+14]
    ;             j2 += incr2;
    mov   ax,word ptr [bp+20]
    add   dx,ax
    inc   si                     ;i
@1@98:
    cmp   si,word ptr [bp+22]
    jle   short @1@50
    ;             *s = tmp;
    FLD   tbyte ptr [bp-10]
    les   bx,dword ptr [bp+6]
    FSTP  qword ptr es:[bx]
    ;             return;
    FWAIT
    pop   di
    pop   si
    leave
    ret
_vdot     endp
```

Table 4.13 Hand coded assembler program for the dot product

```
SPARAM    STRUC
 _BP      DW ?
 PRET     DD ?
 N        DW ?
 INCR2    DW ?
 V2       DD ?
 INCR1    DW ?
 V1       DD ?
 S        DD ?
;pascal vdot(double *s,double v1[],int incr1,double v2[],int
incr2,int n)
VDOT      PROC FAR
   push bp              ; generate stack frame and save registers
used
   mov    bp,sp
   push   ds
   push   si
   push   di
   ;           tmp =0.0e0;
   FLDZ
   ;           j1 = j2 = 0;
   mov    bx,[bp].incr1     ; bx is incr1*8 for 8-byte reals
   sal    bx,3
;
   mov    dx,[bp].incr2     ; dx is incr2*8 for 8-byte reals
   sal    dx,3
;
   lds    si,[bp].v1        ;ds:[si] is pointer to v1
   les    di,[bp].v2        ;es:[di] is pointer to v2
   ;           for (i = 1; i <= n; i++)
LOOP1:
   ;           tmp += v1[j1] * v2[j2];
   FLD    qword ptr [si]
   FMUL   es:qword ptr [di]
   FADD
   ;           j1 += incr1;
   ;           j2 += incr2;
   add    si,bx
   add    di,dx
   loop   loop1
   ;           *s = tmp;
   lds    si,[bp].s
   FSTP   qword ptr [si]
   FWAIT
RETURN:
   pop    di
   pop    si
   pop    ds
   mov    sp,bp
   pop    bp
   ret    12h
VDOT      ENDP
```

In addition to the fast vector arithmetic, the real-time toolbox contains several libraries for real-time control purposes deduced from the standard CADACS control software. This libraries are available in form of Pascal and C sources. Table 4.14 resumes the contents of these libraries. Where possible the arithmetic subroutines rely on the vector instruction set described above. So it is for example possible to express a polynomial multiplication by a call of the VPIV subroutine according to Table 4.10 in a loop. Or the multiplication of two matrices is programmed by a double loop calling the VDOT subroutine.

Table 4.14 Mathematical and control engineering software of the real-time toolbox

Library	Contents
SLINP	The LINPACK library for solution of linear systems of equation for several special forms of the coefficient matrix.
SEISP	The EISPACK library package for matrix eigensystem routines.
MUP	Basic matrix arithmetics for addition, multiplication, elementary and orthogonal transformations etc. of matrices.
PUP	Basic subroutine set for the manipulation of polynomials also including stability test, calculation of roots and spectral factorization of polynomials.
ADAPSISO	Library with numerical subroutines for the implementation of adaptive control algorithms, containing design algorithms for pole placement, model reference, linear quadratic optimal and long-range predictive adaptive controllers.
ADARLS	A library for recursive least-squares and extended least-squares estimation problems with time-varying forgetting, comprising 12 different approaches.
KDIFx	The interfaces to the CADACS data base for transfer functions, signals, matrices.
KED1A	The interface to the CADACS implementation dependent functions as help system, error logging, command input and command interpretation etc.
GRMLB	The graphics subroutine library, providing the interface to the Graphics Manager.
DOKLB	The interface routines for the Documentation Manager for storage and graphical display of real-time data.

Besides the basic routines of matrix and polynomial algebra, the real-time toolbox offers aslo special libraries for the implementation of adaptive control algorithms. ADARLS is used for parameter estimation and ADAPSISO serves for the off and on line design of indirect adaptive controllers for several different control laws. It can also be used for the initial parametrization of direct adaptive control algorithms as for example in case of minimum variance self-tuning controllers or model reference adaptive controllers.

4.6 Conclusion

This chapter introduced the hardware and software for a control system for industrial application and development oriented programming of adaptive control algorithms. The single board computer is usable for the adaptive control of single-input single-output systems in stand-alone operation as well as coupled to a standard PC. The microcomputer is based on a 8086/8087 configuration attached to the EC-bus. For the special requests of automatic control, this microcomputer system has been equipped with interfaces to generate a command level with an intelligent terminal as master and with an interface for high rates for the transportation of data. This data serve for the documentation of process variables and for the test on internal signals and parameters. The state of the control algorithm is steadily monitored by LED's and may be observed by attaching an 8-channel digital data recorder to an 8-pin diagnosis port. A user interface consisting of a teachbox, available for the non-expert operator of the adaptive controller, allows in addition to changing standard settings, e.g. specification of the reference signal, limited influence on the overall performance of the control algorithm. Using the intelligent terminal connection offered by a standard PC, all process specific parameters can be viewed and changed. In addition, a complete documentation of internal parameters and signals is offered for off-line analysis by the program system CADACS.

The development of control engineering software is supported by several special functions of the operating system and by a comprehensive subroutine library in such a way, that no knowledge about the specific hardware of the control equipment is necessary. The hardware and general service software presented in this chapter is not only applicable for the development of controllers for DC-machines, but also for other control purposes and the practical test of control algorithms of several kinds. Proposals for future development are:

Networking. The fast parallel connection is suitable for experiments and test cases but not usable as communication basis in an industrial environment, due to the limited length of the communication line and the high effort for the cabling using 8 data lines in combination with strobe and acknowledge signals. A serial connection involving 3 lines is better suited, but the RS232 technology limits the speed (max. 19200 bit/s) and also the length of the communication line. A standard such as the PROFIBUS or FELDBUS (Halang and Sacha, 1992) is superior to the concepts proposed here and offers, in addition to a fast standard hardware definition, also a state-of-the-art software protocol.

Safety of operation. By additional soft- and hardware this aim can be reached. Non volatile extra RAM which contains all important process and controller data can be used to restore the system after a cold start, e.g. due to power fail, to its nominal state. The check for consistency of input data, the correctness of controller operation and the on line test of hardware operation could be enhanced by additional independent supervisory programs. The watchdog system introduced in the third prototype stage is one step into this direction. The best approach seems to be the use of a second processor, which could be mainly used for the task of interfacing to the user, e.g. operation of the teachbox and the communication protocol. This second processor is free under normal operation conditions, when the control task is operated by the main processor, to perform safety checks and could be used for alarms and emergency support.

4.7 References

Coffron, J.W. (1984),'*Programmierung des 8086/8088*', Sybex Verlag, Düsseldorf.

Digital Research (1983), 'Pascal/MT+, Programmer's Guide for the IBM/PC Disk Operating System', Digital Research Corporation.

Haase, F. (1985), 'Rechnerunterstützte Analyse und Synthese von Regelungsproblemen mit Mikrorechnern', Dr.-Ing. Thesis, Ruhr-Universität Bochum, Germany.

Halang, W.A and K.M. Sacha (1992), '*Real-Time Systems - Implementation of Industrial Computerised Process Automation*', World Scientific Publishers, Singapore.

Intel (1978), 'MCS-48 Microcomputer User's Manual', Intel Corporation, Santa Clara, CA.

Keuchel, U. and Chr. Schmid (1991), 'Identification and control with CADACS-PC', Prepr. of the IFAC Symposium on Identification and System Parameter Estimation, Budapest, pp. 679-684.

Majewski, F. (1989), 'Der EMUF-50', *mc*, Franzis Verlag, München,pp. 123-131.

Mathworks (1989), 'PC-Matlab', The Mathworks Inc., South Natick, MA.

Matrox (1984), 'MBC-86/12B CPU Board', Matrox Electronic Systems Ltd., Manual No. 168-A50-03/1.

Microsoft (1985), 'Microsoft Macro Assembler', Microsoft Corporation.

Microsoft (1991), 'MSDOS Disk Operating System', Microsoft Corporation.

Microsoft (1987), 'Microsoft C User's Manual', Microsoft Corporation.

Microsystems Components Databook (1985), 'Microprocessors and Peripherals I, II', ISBN 0-917017-22-6.

NEC (1988), 'Databook Microprocessors and Peripherals', Part.-No. DBMP-IP...028V20.

Polydata (1983), 'PolyPascal User's and Reference Manual', Polydata, Kopenhagen.

Rash, B. (1981), 'Getting Started with the Numeric Data Processor', Intel Application Note AP-113.

Rector, R. and G. Alexy (1982), *'Das 8086/8088 Buch'*, TE-WI Verlag, München.

Reinsch, C. (1981), 'Der Arithmetik-Prozessor INTEL 8087: ein komplette Implementierung des vorgeschlagenen IEEE-Standards für Gleitpunktarithmetik', *Elektronische Rechenanlagen*, 23, pp. 173-178.

Schmid, Chr. (1985), 'KEDDC - A computer-aided analysis and design package for control systems', in: Jamshidi, M. and C.J. (eds.), *'Computer-Aided Control System Engineering'*, Nort-Holland, Amsterdam, pp.159-180.

Schulman, A., R.J. Brown, J. Kyle, T. Paterson, D. Maxey and R. Brown (1990), *'Undocumented DOS'*, Addison-Wesley, Reading.

Thies, K.D. (1985), *'Die 8087/80287 numerischen Prozessorerweiterungen für 8086/80286 Systeme'*, TE-WI Verlag, München.

Unbehauen, H. (1987) 'KEDDC - Ein Programmsystem zur rechnergestützten Analyse und Synthese von Regelsystemen', *msr*, 18, pp. 78-82.

Unbehauen, H., U. Keuchel and Chr. Schmid (1990) 'Eine Mikrorechner-Arbeitsstation für die rechnergestützte Synthese und Analyse sowie den Echtzeitbetrieb von Regelsystemen', 4. Tagung Elektrotechnologie an der Humbold Universität zu Berlin, Berlin, pp. 233-238.

Werner, K. (1985), 'Der c`t 86 Computer', *c`t*, Heise Verlag Hannover.

CHAPTER 5
BASIC THEORY OF THE IMPLEMENTED ADAPTIVE CONTROL

5.1 Introduction

Adaptive control systems are used for plants, where the dynamic behaviour changes during operation or where the plant parameters are unknown. Simple examples for this type of plants are nonlinear systems, which are run in a certain operating point. A choice of a different operating condition will often lead to changes in the system parameters, which describe the system in the new operating point. More problems in controller design are introduced by steadily changing parameters or changes which are not foreseen. To cope with all possible cases, a perpetual or continuous adaptation must be executed.

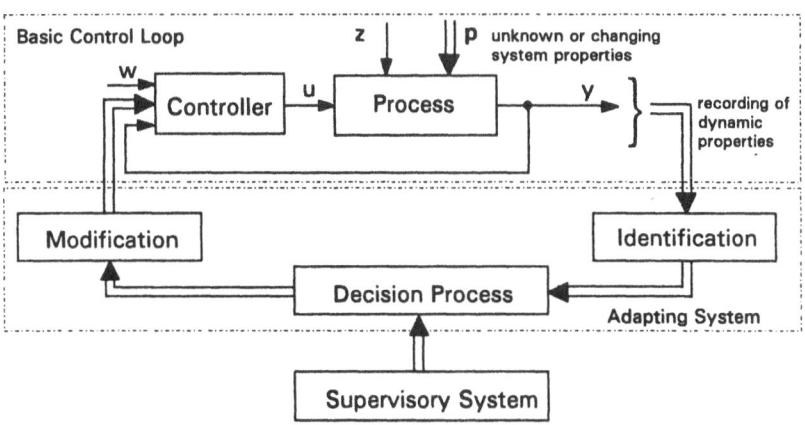

Fig. 5.1 General structure of an adaptive control system

Adaptive control systems can be regarded as an extension of the classical control principles. As illustrated in Fig. 5.1, the basic control loop is superimposed by an adaptation system. Based on the identification, which enables one to ascertain the system properties, the adjustable variables of the controller (parameters, structure etc.) are modified automatically after passing through a decision process. The adapting system and the basic control loop are usually supplemented by a supervisory system for safety purposes. A discussion of the basic terms of adaptive systems can be found in Åström and Wittenmark (1989) or in the standards VDI/VDE GMA 3685 (1990).

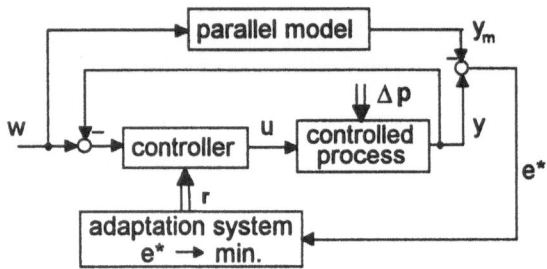

Fig. 5.2 Principle of direct model reference adaptive control

Based on the estimated parameters, the design of the controller is carried out on line. The first control strategy presented in this chapter is based on the model reference principle (Landau, 1979). As depicted in Fig. 5.2, the closed loop system shall behave as specified by a parallel model. The model error e* is fed to an adaptation system which tunes directly the parameters of the controller, such that the error signal e* vanishes or at least will be minimized.

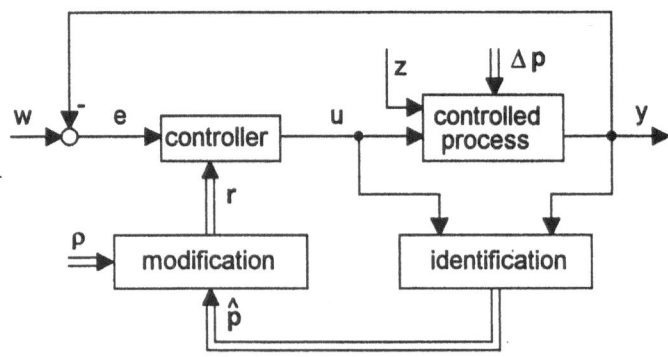

Fig. 5.3 Principle of indirect adaptive control

Fig. 5.3 illustrates the principle of indirect adaptive control which comprises explicit identification of the plant parameters. The second adaptive control structure implemented for speed control of the DC-motor and to be described in section 5.4 is of the indirect type. Its modification stage is based on pole placement design or on a linear quadratic optimal control law.

This chapter will present the basic theory for the adaptive control strategies implemented in the control instrumentation. Furthermore, implementation aspects necessary for the reliable robust and stable behaviour of the adaptive controller under practical operating conditions will be discussed.

5.2 The Model Reference Adaptive Controller

5.2.1 Basic Considerations

The model of the plant in z-domain notation with single input $U(z)$, single output $Y(z)$ and unit delay is described by the discrete transfer function

$$G_s(z) = \frac{B(z)}{A(z)} \, z^{-1} = \frac{Y(z)}{U(z)} \tag{5.1}$$

with polynomials

$$A(z) = 1 + (a_1 + a_2 z^{-1} + \ldots + a_n z^{-(n-1)}) \, z^{-1} = 1 + A^*(z) \, z^{-1} \tag{5.2}$$

$$B(z) = b_0 + (b_1 + b_2 z^{-1} + \ldots + b_n z^{-(n-1)}) \, z^{-1} = b_0 + B^*(z) \, z^{-1}. \tag{5.3}$$

For simplicity of treatment, only the unit delay case is considered first. However, the results can be generalized for plants with arbitrary time delays as it will be shown in section 5.2 and for systems with multiple inputs and multiple outputs (Hahn, 1983; Unbehauen and Wiemer, 1985). The disturbed plant output is described by

$$A(z) \, Y(z) = B(z) \, z^{-1} U(z) + V(z) \tag{5.4}$$

$$V(z) = C(z) \, \varepsilon(z) + Z_d(z) \tag{5.5}$$

where $\varepsilon(z)$ is an independent white noise signal and $Z_d(z)$ denotes a non measurable deterministic disturbance.

In order to stabilize the nonminimum phase control, the idea of the *correction network* introduced by Hahn and Unbehauen (1982) is applied. The plant output $Y(z)$ is, therefore, augmented by the signal

$$Y_c(z) = G_c(z) \, U(z) \tag{5.6}$$

with

$$G_c(z) = \frac{B_c(z)}{A_c(z)} z^{-1} = \frac{Y_c(z)}{U(z)}$$ (5.7)

and

$$A_c(z) = 1 + A_c^*(z)z^{-1}$$ (5.8)

$$B_c(z) = b_{c0} + B_c^*(z)z^{-1}$$ (5.9)

wherein the correction network $G_c(z)$ is supposed to be stable. The augmented plant output is then

$$Y_a(z) = Y(z) + Y_c(z) .$$ (5.10)

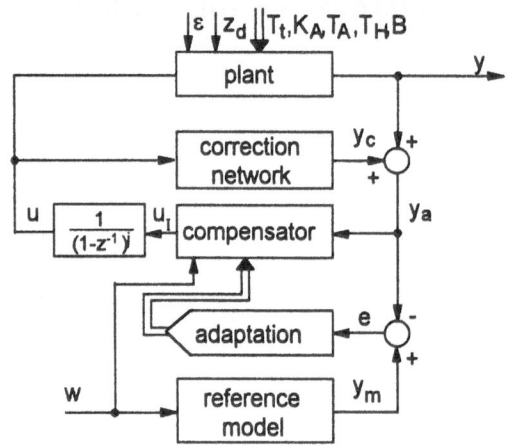

Fig. 5.4 Basic structure of the implemented MRAC scheme

In order to reject the influence of deterministic disturbances which are assumed to have the properties

$$\lim_{k \to \infty} z_d(k) = \text{const,}$$ (5.11)

it is convenient to force the controller to have integral action. This can easily be obtained by multiplying Eqs. (5.4) and (5.7) with the factor $(1-z^{-1})$ on both sides. Therefore, they are modified to

$$A_I(z) Y(z) = B(z)z^{-1}U_I(z) + V_I(z)$$ (5.12a)

and

$$A_{cl}(z) Y_c(z) = B_c(z)z^{-1}U_I(z)$$ (5.12b)

with

$$A_I(z) = (1-z^{-1})^j A(z) = 1 + A_I^*(z)z^{-1}$$ (5.13a)

70

$$U_1(z) = (1-z^{-1})^j\, U(z) \tag{5.13b}$$

$$V_1(z) = (1-z^{-1})^j\, V(z) \tag{5.13c}$$

$$A_{cl}(z) = (1-z^{-1})^j\, A_c(z) = 1 + A_{cl}^*(z)z^{-1} \tag{5.13d}$$

$$j \in \{0, 1\}. \tag{5.13e}$$

Note that by this modification and under the assumption of Eq. (5.11) and $\varepsilon(z)=0$

$$\lim_{k\to\infty} v_1(k) = 0, \text{ for } j = 1. \tag{5.14}$$

This basic structure of the control scheme is shown in Fig. 5.4.

5.2.2 The Control Law

The augmented plant output $Y_a(z)$ is compared with the output $Y_m(z)$ of the stable reference model

$$Y_m(z) = G_m(z)\, W(z) \tag{5.15}$$

where

$$G_m(z) = \frac{B_m(z)}{A_m(z)}\, z^{-1} = \frac{Y_m(z)}{W(z)} \tag{5.16}$$

and

$$A_m(z) = 1 + A_m^*(z)z^{-1}, \tag{5.17}$$

wherein $W(z)$ is the reference input (set point).

The dynamic behaviour of the error signal

$$E(z) = Y_m(z) - Y_a(z) \tag{5.18}$$

can be expressed by

$$A_m(z)\, E(z) = A_m(z)\, Y_m(z) - A_m(z)\, Y(z) - A_m(z)\, Y_c(z). \tag{5.19}$$

Adding

$$[A_1(z)\, Y(z) - B(z)z^{-1}U_1(z) - V_1(z)] + [A_{cl}(z)\, Y_c(z) - B_c(z)z^{-1}U_1(z)\,] = 0$$

to the right hand term of Eq. (5.19) and recalling from Eq. (5.16) that

$$A_m(z)\, Y_m(z) = B_m(z)z^{-1}\, W(z)$$

it follows that

$$A_m(z)\, E(z) = z^{-1}\{B_m(z)W(z) + [A_1^*(z) - A_m^*(z)]\, Y(z) + [A_{cl}^*(z) - A_m^*(z)]\, Y_c(z) \\ - [B(z) + B_c(z)]\, U_1(z)\,\} - V_1(z). \tag{5.20}$$

Now, using Eqs. (5.3) and (5.9) the error difference equation, Eq. (5.20), obtains the form

$$E_m(z) \overset{!}{=} A_m(z) \, E(z) = z^{-1}\{ \, R(z)-[b_0+b_{c0}]U_1(z)-B^*(z)z^{-1}U_1(z)+\Delta A^*(z)Y(z)\}- V_1(z)$$
$$(5.21)$$

with

$$\Delta A^*(z) = A_1^*(z) - A_m^*(z) \tag{5.22}$$

and

$$R(z) = B_m(z) \, W(z) - B_c^*(z)z^{-1}U_1(z) + [\, A_{cl}^*(z) - A_m^*(z)] \, Y_c(z) \,. \tag{5.23}$$

The signal $R(z)$ is generated only with known parameters and measurable plant input and output signals.

If the manipulated signal is computed by

$$U_1(z) = \frac{1}{b_0 + b_{c0}} [R(z) - B^*(z)z^{-1}U_1(z) + \Delta A^*(z)Y(z)] \,, \tag{5.24}$$

for $V_1(z)=0$ the augmented plant output $Y_a(z)$ follows the output of the reference model $Y_m(z)$, i.e. $E_m(z)=0$.

Note that the polynomial $B_c(z)$ is forced to have a zero at $z=1$ if integral action is required. Then, under the assumption

$$\lim_{k\to\infty} u(k) \; = \; const, \tag{5.25}$$

the output $y_c(k)$ of the correction network vanishes for $k\to\infty$. Eq. (5.25) can only be satisfied for constant reference signals $W(z)$ and step disturbances $Z_d(z)$. In this case we have

$$\lim_{k\to\infty} y(k) = \lim_{k\to\infty} y_m(k), \tag{5.26}$$

i.e. asymptotic matching of the plant model output.

In Eq. (5.24) the coefficients of the polynomials $B^*(z)$ and $\Delta A^*(z)$ depend on the unknown plant parameters. For the computation of the adaptive control law they have to be replaced by their estimates

$$U_1(z) = \frac{1}{\hat{b}_0 + b_{c0}} [R(z) - \hat{B}^*(z)z^{-1}U_1(z) + \Delta \hat{A}^*(z)Y(z)] \,. \tag{5.27}$$

In Eq. (5.27) this is indicated by the symbol "^". This equation can easily be solved, provided the term $\hat{b}_0 + b_{c0}$ does not vanish. This can always be achieved by choosing b_{c0} for the correction network in an appropriate way. With Eq. (5.13b) the output of the adaptive controller is

$$U(z) = \frac{1}{(1-z^{-1})^j} U_1(z) \quad , \text{ with } j \in \{0,1\}. \tag{5.28}$$

5.2.3 Adaptation of Parameters

By substituting R(z) from Eq. (5.27) in Eq. (5.21) we obtain

$$E_m(z) = A_m(z) E(z) \tag{5.29}$$
$$= z^{-1}\{[\hat{b}_0 - b_0]U_1(z) + [\hat{B}^*(z)-B^*(z)]z^{-1}U_1(z) - [\Delta\hat{A}^*(z)-\Delta A^*(z)]Y(z)\}-V_1(z).$$

Considering now time varying controller parameters Eq. (5.29) can be rewritten as

$$e_m(k) = [p - \hat{p}(k-1)]^T m(k-1)-v_1(k), \tag{5.30}$$

where all signals are contained in the *signal vector* m(k) and all parameters are included into the *parameter vectors* p and $\hat{p}(k)$. For the on-line computation of Eq. (5.27), it is necessary to have a short delay between measurement of the plant output and output of the new control signal. Therefore in Eq. (5.27) the parameters estimated at the previous step (k-1) should be used. Then Eq. (5.30) has to be modified as

$$e_m(k) = [p - \hat{p}(k-2)]^T m(k-1)-v_1(k) . \tag{5.31}$$

We define an *augmented error signal*

$$\dot{e}(k) = e_m(k) + h(k) \tag{5.32}$$

where

$$h(k) = [\hat{p}(k-2) - \hat{p}(k-1)]^T m(k-1) . \tag{5.33}$$

This yields

$$\dot{e}(k) = [p - \hat{p}(k-1)]^T m(k-1)-v_1(k) . \tag{5.34}$$

Eq. (5.34) describes an estimation problem, which is very well studied at least in the disturbance free and white noise case (Lozano and Landau, 1981). This estimation problem can be formulated as follows. Find a law for the adaptation of the controller parameters $\hat{p}$ such that

$$\lim_{k\to\infty} \dot{e}(k) = 0. \tag{5.35}$$

There are different solutions available for this estimation problem. As a least squares approach takes more computational effort, a *generalized stochastic approximation method* (Hahn, 1983), similar to the algorithm proposed by (Goodwin et al., 1981), can be applied for the estimation of the controller parameters. In this case the parameter update yields

$$\hat{p}(k) = \hat{p}(k-1) + r(k) G m(k-1) \dot{e}(k) \tag{5.36}$$

where

$$r(k) = \begin{cases} r^*(k) & \text{if } r^*(k) \geq \dfrac{1}{\rho\, \mathbf{m}^T(k\text{-}1)\, \mathbf{G}\, \mathbf{m}(k\text{-}1) + \gamma} \\ \text{or} \\ \dfrac{1}{\rho\, \mathbf{m}^T(k\text{-}1)\, \mathbf{G}\, \mathbf{m}(k\text{-}1) + \gamma} & \text{otherwise} \end{cases} \qquad (5.37)$$

$$\frac{1}{r^*(k)} = \frac{\lambda_1(k)}{r(k\text{-}1)} + \lambda_2(k)\, \mathbf{m}^T(k\text{-}1)\, \mathbf{G}\, \mathbf{m}(k\text{-}1) + \lambda_3(k) \qquad (5.38)$$

and the values of

$$G > 0 \qquad , r(\text{-}1) > 0 \qquad\qquad\qquad\qquad (5.39a)$$

$$0 \leq \lambda_1(k) \quad ; \frac{1}{2} < \lambda_2(k) \qquad\qquad\qquad\qquad (5.39b)$$

$$0 \leq \lambda_3(k) \quad ; \rho > \frac{1}{2} \quad ; \gamma \geq 0 \qquad\qquad\qquad (5.39c)$$

are freely selectable. This algorithm, Eqs. (5.36-5.39), contains the "classical" adaptation algorithm with non decreasing (fixed) gain estimation (Ionescu and Monopoli, 1977) as a special case, setting

$$\lambda_1 = 0;\ \lambda_2 > 0.5;\ \lambda_3 = 1\,. \qquad\qquad\qquad (5.40a)$$

The stability of the total algorithm in the disturbance free case, i.e. $v(k)\equiv0$ is guaranteed if $G_c(z)$ and $G_m(z)$ are stable and the parallel connection $G_s(z)+G_c(z)$ has no zeros for $|z|\geq1$. Then all signals are bounded. The algorithm can be extended to the disturbed case by introducing a "dead zone" into the adaptation law (Peterson and Narendra, 1982). For the proof of stability see (Hahn, 1983). As the reference model is stable, the convergence of the original error signal $e(k)$ directly follows. The influence of step disturbances on the estimation procedure is at least asymptotically rejected if an explicit integrator $(j=1)$ is used because then Eq. (5.14) holds.

5.3 The Improved Model Reference Adaptive Controller

5.3.1 Problems in Standard Model Reference Adaptive Control

This section deals with a discrete-time robust adaptive control strategy for single-input single-output systems with arbitrary zeros, which is a generalization of the control strategy proposed in section 5.1. The basic idea presented here is to divide the feedback law into a direct adaptive and an indirect adaptive part. By a certain choice of these indirectly synthesized adaptive controller parameters the robustness

74

of the control loop with respect to unmodelled dynamics of the plant can be increased. The indirect adaptive part of the controller is just the "correction network" introduced in the last section, but it will be discussed here from a different point of view. The modified control strategy removes the major drawbacks of model reference control.

In recent years several publications stressed the robustness properties of adaptive control algorithms. They stated, that for example standard model reference adaptive control (Monopoli, 1974; Lozano and Landau, 1981) tends in practice to instability due to highly oscillatory modes, despite design according to stability theory. But Rohrs and Shortelle (1984) also illustrated, that a special bypass to the plant can solve the problem. Wiemer and Unbehauen (1990) presented a stability theorem, where a special filter called "correction network" was used to increase robustness with respect to unmodelled interconnections between the different control loops in decentralized adaptive control.

Former approaches (Hahn and Unbehauen, 1982), as discussed in section 5.1, used fixed correction networks, which had to be synthesized off-line based on a priori information about the plant. The major drawback of this strategy is, that variations of the plant parameters can only be taken into account by a robust design for the correction network. To overcome these difficulties an on-line update of the correction network based on estimated plant parameters is suggested here. The design procedure leads to a simple pole placement problem, such that the desired performance of the closed-loop system can be characterized by a few well-understood parameters. Closed-loop pole positions may either be specified in terms of damping and bandwidth, or by the weighting factor for the manipulating signal for infinite horizon linear quadratic optimal control. The optimal characteristic polynomial is evaluated by on-line solution of a spectral factorization problem.

Combination of the direct adaptive model reference scheme with on-line synthesis of the correction network, together with fast and stable algorithms for estimation of plant and controller parameters, results in an adaptive control algorithm which is easy to implement and simple to handle (Unbehauen, 1989). After introducing the linear control loop, an adaptive control law will be derived. The next section discusses the design of the correction network.

5.3.2 The Linear Control Law

The structure of the linear control loop is depicted in Fig. 5.5. For description of plant and controller discrete transfer functions have been chosen. It should be noted, that controller polynomials R, S and T as well as the polynomials A and B of the plant transfer function

$$G(z) = \frac{Y(z)}{U(z)} = \frac{B(z)}{A(z)} z^{-d} \tag{5.41}$$

with dead time $T_t = dT$ (T: sampling time) are polynomials in z^{-1}. The actuating signal u(k) is generated by discrete integration from $u_l(k)$. The integral term at the controller output implements a switchable integral control law. This modification has been provided especially for rejection of deterministic disturbances.

For setpoint behaviour, neglecting all disturbances, the control law in z-domain is represented by

$$U(z) = [T(z) W(z) - S(z) Y(z)] \frac{1}{R(z)(1-z^{-1})^j}, \quad j \in \{0,1\}. \tag{5.42}$$

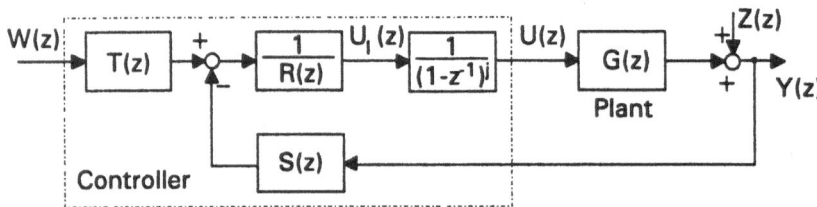

Fig. 5.5 The linear control loop

For design of the controller polynomials it will be required that the setpoint behaviour shall correspond to a prespecified model behaviour, i.e.

$$G_w(z) \overset{!}{=} G_m(z) = \frac{Y_m(z)}{W(z)} = \frac{B_m(z)}{A_m(z)} z^{-d}, \tag{5.43}$$

where $G_m(z)$ represents the transfer function of a reference model. Substitution of $P(z)=R(z)B^{-1}(z)$ into the setpoint transfer function

$$G_w(z) = \frac{Y(z)}{W(z)} = \frac{T(z) z^{-d}}{(1-z^{-1})^j R(z) B^{-1}(z) A(z) + S(z) z^{-d}}, \tag{5.44}$$

leads in combination with Eq. (5.43) to the identities

$$B_m(z) = T(z) \tag{5.45a}$$

$$A_m(z) = (1-z^{-1})^j P(z) A(z) + S(z) z^{-d}. \tag{5.45b}$$

Eqs. (5.45 a,b) are used for synthesis of the controller. As is obvious, these design equations are only applicable to minimum phase plants due to cancellation of the plant numerator polynomial B(z) in Eq. (5.44). If B(z) has roots outside the unit circle of the complex z-plane, i.e. if it is nonminimum phase, a cancellation may not occur due to stability reasons. From the viewpoint of generalization with respect to practical demands (many industrial plants exhibit nonminimum phase behaviour or become nonminimum phase after discretization) these important cases should be included into the design procedure.

A well proved and efficient method to solve this problem is the introduction of a correction network (Hahn and Unbehauen, 1982). Fig. 5.6 shows the augmented structure of the control loop in combination with the correction network having transfer function

$$G_c(z) = \frac{Y_c(z)}{U_l(z)} = \frac{B_c(z)}{A_c(z)} z^{-d} \qquad (5.46)$$

with

$$B_c(z) = (1-z^{-1})^{1-j} B_c'(z), \quad j \in \{0,1\} . \qquad (5.47)$$

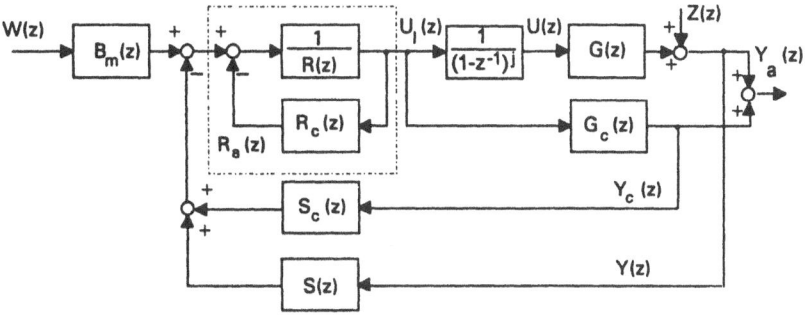

Fig. 5.6 The augmented linear control loop

As Eq. (5.47) illustrates, the correction network exhibits differential behaviour, which will only be switched on, if no integral control law was specified (j=0). Thus the steady state control deviation introduced by the correction network will be avoided. To guarantee the model behaviour, according Eq. (5.43), also for the augmented plant one requests

$$Y_a(z) = [\, G_c(z) + (1-z^{-1})^{-j} G(z) \,] \, U_l(z) \stackrel{!}{=} Y_m(z) .$$

The transfer function of the augmented system from input $U_l(z)$ to output $Y_a(z)$ is given by

$$G_a(z) = \frac{Y_a(z)}{U_l(z)} = \frac{A_c(z)B(z)+(1-z^{-1})A(z)B_c(z)}{A(z)A_c(z)(1-z^{-1})^j} z^{-d} = \frac{B_a(z)}{A_a(z)} z^{-d} . \qquad (5.48)$$

The steady-state behaviour of the correction network is governed by

$$\lim_{k\to\infty} y_c(k) = 0,$$

and due to the choice of the numerator polynomial

$$\lim_{k\to\infty} y(k) = \lim_{k\to\infty} y_a(k) = \lim_{k\to\infty} y_m(k).$$

According to Eq. (5.48) the augmented plant will be of order

$$\deg A_a = \deg A + \deg A_c + j. \tag{5.49}$$

If a controller following Eq. (5.45) is for the augmented plant, the degree of controller polynomials would grow in an undesirable way due to the increased order of the augmented plant. With respect to the adaptive version of the control algorithm to be discussed in the next section, the controller polynomial R_a will be split according to

$$R_a(z) = R(z) + R_c(z) = P(z) B(z) + P_c(z) B_c(z) \tag{5.50a}$$

with

$$P_c(z) = R_c(z) B_c^{-1}(z). \tag{5.50b}$$

Thus the signals $y(k)$ and $y_c(k)$ will be fed back separately using controller polynomials $S(z)$ and $S_c(z)$, which is illustrated in Fig. 5.7. Thus the control law

$$U_1(z) = \frac{1}{R_a(z)} [B_m(z) W(z) - S(z) Y(z) - S_c(z) Y_c(z)] \tag{5.51}$$

is obtained. The setpoint transfer function of the augmented control loop is given by

$$G_{wa}(z) = \frac{Y_a(z)}{W(z)} = \frac{[(1-z^{-1})^j B_c(z)A(z)+A_c(z)B(z)]z^{-d}}{(1-z^{-1})^j A(z)[A_c(z)R_a(z)+S_c(z)B_c(z)z^{-d}]+S(z)A_c(z)B(z)z^{-d}}. \tag{5.52}$$

For model following the design equations for the augmented controller are derived from

$$A_m(z)[(1-z^{-1})^j B_c(z)A(z)+A_c(z)B(z)] = (1-z^{-1})^j A(z)[A_c(z)R_a(z)+S_c(z)B_c(z)z^{-d}]+S(z)A_c(z)B(z)z^{-d}. \tag{5.53}$$

By substituting R_a from Eq. (5.50a,b) and taking into account Eq. (5.45a,b) one obtains

$$A_m(z)= P_c(z)A_c(z)+ S_c(z)z^{-d} \tag{5.54}$$

as the design equation for the second part of the controller. So the controller polynomials for the augmented system may be evaluated as follows:

a) T from Eq. (5.45a),

b) P, respectively R and S from Eq. (5.45b),

c) P_c, respectively R_c and S_c from Eq. (5.54).

5.3.3 Design of the Correction Network

By introduction of the correction network $G_c(z)$ the minimum phase characteristics of the augmented plant has to be guaranteed. Therefore the roots of the numerator polynomial of $G_a(z)$, Eq. (5.48),

$$B_a(z) = A_c(z) B(z) + (1-z^{-1})^j A(z) B_c(z) \qquad (5.55)$$

have to be located within the unit circle of the z-plane.

The design of the correction network can be performed in several ways. Former approaches (Hahn and Unbehauen, 1982) used a fixed correction network, which was not tuned on-line to compensate for changes in plant parameters. Using root-locus plots the polynomials A_c and B_c had to be designed off-line, starting from a raw plant model, such that minimum phase behaviour for the augmented plant could be reached even in the case of parameter variations. This method was effective and quite simple to use for the skilled design engineer, who could influence the closed-loop and actuation dynamics by a few variable parameters, such as prefactor or time constants of the correction network. But this approach turned out to be quite difficult for the non-experienced user of the adaptive control algorithm.

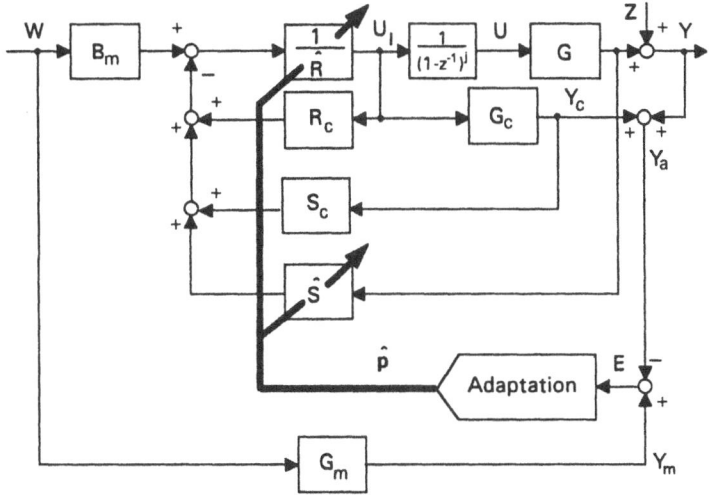

Fig. 5.7 The adaptive model reference control loop

The tuning procedure for the correction network is based on estimated plant parameters. Therefore additional on-line identification for plant parameters $A(z)$ and $B(z)$ is necessary, since recalculation from estimated controller parameters is not possible. The tracking behaviour of the whole control loop, Fig. 5.7, with a well adjusted correction network and controller polynomials is given by

$$Y(z) = \frac{B(z)\, A_c(z)\, B_m(z)}{B_a(z)\, A_m(z)}\, z^{-d}\, W(z). \tag{5.56}$$

The closed loop poles are given by the roots of $A_m(z)$ and the numerator roots of the augmented plant. The design of the correction network according to Eq. (5.55) then results in a simple pole placement problem.

To simplify the operation of the adaptive controller, the desired performance of the closed-loop system should be characterized by a few well-understood parameters. In the implemented adaptive control scheme, system behaviour may either be specified in terms of relative damping d and bandwidth ω_0 leading to

$$B_a(z) = 1 + b_{a1} z^{-1} + b_{a2} z^{-2}, \tag{5.57}$$

with

$$b_{a1} = -2e^{d\omega_0 T} \cos(\omega_0 T)\,(\,1\text{-}d^2) \quad \text{and} \quad b_{a2} = e^{-2d\omega_0 T}, \tag{5.58}$$

or by a linear quadratic control strategy which minimizes the cost function

$$\sum_{k=0}^{\infty} (y^2(k) + \rho\, u^2(k)\,) \overset{!}{=} \min. \tag{5.59}$$

In the last case the optimal closed loop characteristic polynomial (for $A_m=B_m=1$) is obtained by solving the spectral factorization problem

$$B_a(z)\, \Gamma\, B_a(z^{-1}) = \rho\, A(z)\, A(z^{-1}) + B(z)\, B(z^{-1}). \tag{5.60}$$

By choosing d, ω_0 or ρ, the minimum phase assumption for the augmented plant can be fulfilled, additionally the closed loop dynamics for the controlled value can be specified.

5.3.4 Calculation of the Filtered Model Error

If a non measurable deterministic disturbance $Z(z)$ at the output of the plant is introduced, the controlled signal is given by

$$Y(z) = \frac{B(z)}{A(z)}\, z^{-d}\, \frac{1}{(1\text{-}z^{-1})^j}\, U_l(z) + Z(z). \tag{5.61}$$

To derive an expression for the manipulated signal, instead of the model error the filtered error

$$E_m(z) = A_m(z)E(z) = A_m(z)Y_m(z) - A_m(z)Y(z) - A_m(z)Y_c(z) \qquad (5.62)$$

is used, which may also be written as

$$E_m(z) = B_m(z)z^{-d}W(z) - [(1-z^{-1})^j P(z) A(z) + S(z) z^{-d}] Y(z)$$
$$- [P_c(z)A_c(z) + S_c(z)z^{-d}] Y_c(z). \qquad (5.63)$$

Using Eqs. (5.46), (5.50a), (5.61), Eq. (5.63) may be simplified to

$$z^d E_m(z) = B_m(z)W(z) - [R(z) + R_c(z)]U_1(z) - S(z) Y(z) - S_c(z)Y_c(z)$$
$$- P(z) A(z) Z(z) (1-z^{-1})^j z^d. \qquad (5.64)$$

For an asymptotically vanishing filtered error, the left hand side of Eq. (5.64) must be set equal to zero, so that the control law

$$U_1(z) = \frac{1}{r_0+r_{c0}} \{B_m(z)W(z)-[R^*(z)+R_c^*(z)]z^{-1}U_1(z)-S(z)Y(z)-S_c(z)Y_c(z)$$
$$-P(z)A(z)Z(z)(1-z^{-1})^j z^d\} \qquad (5.65)$$

with

$$R = r_0 + R^*(z)z^{-1}$$

and

$$R_c(z) = r_{c0} + R_c^*(z) z^{-1}$$

follows.

If the deterministic disturbance is constrained to be a step

$$Z(z) = \frac{z}{z-1} v_o, \qquad (5.66)$$

the part of the control law, Eq. (5.65), to asymptotically compensate this disturbance is derived from

$$\lim_{z\to\infty}\{ (z-1) P(z)A(z)\frac{z}{z-1} v_0(1-z^{-1})^j z^d\} = (1-j) c, \qquad (5.67)$$

where

$$c = v_0 \sum_{v=0}^{n_A} [p_v \sum_{\mu=0}^{n_P} a_\mu].$$

From Eq. (5.67) it is obvious, that by an integral control law the step disturbance will be rejected asymptotically. The control law in time domain is given by

$$u_1(k) = \frac{1}{r_0+r_{c0}} [\hat{p}_1^T m_1(k) + \hat{p}_{01}^T m_{01}(k)] \qquad (5.68)$$

with parameter vectors

$$\hat{p}_1^T = - [\, \hat{r}_1 \ldots \hat{r}_{nr} \,|\, \hat{s}_0 \ldots \hat{s}_{ns} \,|\, (1\text{-}j) \, \hat{c} \,] \tag{5.69a}$$

$$\hat{p}_{01}^T = - [\, \hat{r}_{cl} \ldots \hat{r}_{cnrc} \,|\, \hat{s}_{c0} \ldots \hat{s}_{cnsc} \,|\, b_{m0} \ldots b_{mnbm} \,] \tag{5.69b}$$

and signal vectors

$$m_1(k) = [u_1(k\text{-}1) \ldots u_1(k\text{-}n_r) \,|\, y(k) \ldots y(k\text{-}n_s) \,|\, 1 \,]^T \tag{5.69c}$$

$$m_{01}(k) = [u_1(k\text{-}1) \ldots u_1(k\text{-}n_{rc}) \,|\, y_c(k) \ldots y_c(k\text{-}n_{sc}) \,|\, \text{-}w(k) \ldots \text{-}w(k\text{-}n_{bm}) \,]^T \tag{5.69d}$$

The coefficients of $\hat{p}_{01}$ have to be calculated from estimated plant parameters $\hat{B}$ and $\hat{A}$ according to Eq. (5.55). As in standard model reference adaptive control the parameters of $\hat{p}_1$ may be estimated directly. Therefore the control law in frequency domain is represented as

$$U_1(z)(\hat{r}_0 + \hat{r}_{c0}) = \hat{M}^*(z) - \hat{R}_c^*(z) \, z^{-1} \, U_1(z) - \hat{S}(z) \, Y(z) - (1\text{-}j) \, \hat{c}(z) \tag{5.70}$$

with

$$\hat{M}^*(z) = B_m(z)W(z) - \hat{R}_c^*(z) \, z^{-1} U_1(z) - \hat{S}_c(z)Y_c(z).$$

If Eq. (5.70) is rewritten as

$$\hat{M}^*(z) = U_1(z) \, (\hat{r}_0 + \hat{r}_{c0}) + \hat{R}^*(z)z^{-1} U_1(z) + \hat{S}(z) \, Y(z) + (1\text{-}j) \, \hat{c}(z), \tag{5.71}$$

then, in combination with Eq. (5.67), follows for the filtered error

$$E_m(z) = z^{-d} \{[\, \hat{R}(z) - R(z)] \, U_1(z) + (\, \hat{S}(z) - S(z)) \, Y(z) \} + (1\text{-}j) \, [\, \hat{c}(z) - c(z) \,] \tag{5.72}$$

which may be represented in time domain as

$$e_m(k) = [p - \hat{p}^T(k\text{-}1)] \, m(k\text{-}d) \tag{5.73}$$

with

$$p^T = - [\, r_0 \ldots r_{nr} \,|\, s_0 \ldots s_{ns} \,|\, (1\text{-}j) \, c \,] \tag{5.74a}$$

$$\hat{p}^T(k\text{-}1) = - [\, \hat{r}_0(k\text{-}1) \ldots \hat{r}_{nr}(k\text{-}1) \,|\, \hat{s}_0(k\text{-}1) \ldots \hat{s}_{ns}(k\text{-}1) \,|\, (1\text{-}j) \, \hat{c}(k\text{-}1) \tag{5.74b}$$

$$m(k\text{-}d) = [\, u_1(k\text{-}d) \ldots u_1(k\text{-}d\text{-}n_r) \,|\, y(k\text{-}d) \ldots y(k\text{-}d\text{-}n_s) \,|\, 1 \,]^T. \tag{5.74c}$$

Eq. (5.73) can only be evaluated for known "true" controller parameters, but by substituting Eq. (5.61) and Eq. (5.67), the terms depending on these parameters may be eliminated. So the filtered error may be calculated from

$$E_m(z) = \hat{R}(z) \, z^{-d} U_1(z) + \hat{S}(z) \, z^{-d} Y(z) + (1\text{-}j) \, \hat{c}(z) - A_m(z)Y(z). \tag{5.75}$$

By inverse z-transformation Eq. (5.75) may be represented in time domain as

$$e_m(k) = a_m^T \, y(k) - \hat{p}^T(k\text{-}1) \, m(k\text{-}d) \tag{5.76}$$

82

with

$$\mathbf{a_m^T} = [\, a_{m0} \;\; \cdots \;\; a_{mn_{am}} \,] , \tag{5.77a}$$

$$\mathbf{y(k)} = [\, y(k) \;\; \cdots \;\; y(k-n_{am}) \,]^T . \tag{5.77b}$$

5.3.5 Estimation Procedures for Plant and Controller Parameters

Estimation of plant and controller parameters is done according to a recursive least squares method

$$\hat{p}(k) = \hat{p}(k-1) + q(k)\,\varepsilon(k) \tag{5.78}$$

where $\varepsilon(k)$ is either the prediction error or the filtered error and $q(k)$ is a Kalman gain vector. The algorithms actually implemented are a RLS-algorithm with σ-modification and variable forgetting factor (Wiemer, 1988) and a square-root filter algorithm with constant trace of the covariance matrix (Unbehauen, 1989), which gave also good performance and results in previous applications. Only the last algorithm was used throughout the experiments, and will be shortly discussed here.

Generally, the update of the Kalman gain vector may be written as

$$q(k) = \frac{P(k-1)\,m(k-d)}{1 + m^T(k-d)\,P(k-1)\,m(k-d)} \tag{5.79}$$

and the covariance matrix of the current time step is calculated from

$$P(k) = \frac{1}{\rho(k)}\,[\, I - q(k)\,m^T(k-d) \,]\,P(k-1) , \tag{5.80}$$

where $0<\rho(k)\leq 1$ is a weighting factor, which weights the measurements according to

$$P^{-1}(k) = \rho(k)\,P^{-1}(k-1) + \rho(k)\,m(k-d)\,m^T(k-d) . \tag{5.81}$$

To guarantee the positive definiteness of $P(k)$ the actual computation is done in terms of the square root $S(k)$ of the covariance matrix (Bierman, 1977)

$$P(k) = S(k)\,S^T(k) . \tag{5.82}$$

Using the factorized form of the covariance matrix, the recursive least squares algorithm may be rewritten as

$$\hat{p}(k) = \hat{p}(k-1) + \frac{g(k-1)}{h^2(k-1)}\,\varepsilon(k) \tag{5.83}$$

with

$$g(k-1) = S(k-1)\,f(k-1) , \quad f(k-1) = S^T(k-1)\,m(k-d) ,$$

$$h^2 = 1 + m^T(k\text{-}d) \, S(k\text{-}1) \, S^T(k\text{-}1) \, m(k\text{-}d) \,.$$

Using the above abbreviations, the square root covariance update yields

$$S(k) = \frac{1}{\sqrt{\rho(k)}} \, S(k\text{-}1) \, [\, I - \frac{1}{h^2 + h} \, f(k\text{-}1) \, f^T(k\text{-}1) \,] \,. \tag{5.84}$$

The requirement of constant trace of the covariance matrix

$$\mathrm{tr} \, (\, S(k) \, S^T(k) \,) = \mathrm{tr} \, P \, (k) = \mathrm{tr} \, P(k\text{-}1) = \mathrm{tr} \, P(0) \tag{5.85}$$

can be fulfilled, if a weighting factor

$$\rho(k) = 1 - \frac{1}{\mathrm{tr} \, S(0) S^T(0)} \, \frac{g^T(k\text{-}1) \, g(k\text{-}1)}{h^2} \tag{5.86}$$

is chosen. But the evaluation of the weighting factor according to Eq. (5.85) will lead in practice to numerical instability of the estimation algorithm. Due to rounding errors it is possible that $\rho(k) > 1$ is calculated. To avoid this problem it is necessary to calculate first

$$S'(k) = \, S(k\text{-}1) \, [\, I - \frac{1}{h^2 + h} \, f(k\text{-}1) \, f^T(k\text{-}1) \,] \tag{5.87}$$

and to evaluate the weighting factor as

$$\rho(k) = \frac{\mathrm{tr} \, (\, S'(k) \, S^T(k) \,)}{\mathrm{tr} \, (\, S(0) \, S^T(0) \,)} \,. \tag{5.88}$$

The additional arithmetic operations for evaluation of Eq. (5.88) can be arranged in an efficient way, if the lower triangular matrix $S'=[s_{ij}]$ is stored in packed form as the one dimensional vector

$$s = [\, s_{11} \quad s_{21} \quad s_{22} \quad s_{31} \quad s_{32} \quad s_{33} \quad \dots \quad s_{nn} \,]^T \,.$$

In this case the trace of the unweighted covariance matrix may be evaluated by simply calculating a scalar product according to

$$\mathrm{tr} \, (\, S'(k) \, S^T(k) \,) = s^T(k) \, s(k) \,. \tag{5.89}$$

Besides initial values for the parameters to be estimated $\hat{p}(0) = p_0$, only one parameter $\mathrm{tr}(P)$ has to be specified. The trace of P is the sum of the eigenvalues of the covariance matrix and thus an upper bound for the largest eigenvalue of this positive definite matrix. The choice of $\mathrm{tr} \, P(0)$ is based on the noise level, on the estimated variation of parameters and on the structural uncertainty. A good guess for $P(0)$ may be supported by the analysis methods derived by Wiemer and Unbehauen (1990).

This section presented a modified adaptive control algorithm according to the model reference principle for single-input single-output systems. The improved control strategy can also handle nonminimum phase systems, if a parallel filter is connected to the plant. An automatic tuning procedure for that filter, which is updated adaptively based on estimated plant parameters was suggested. Estimation procedures for the plant parameters, which are needed for the design of the correction network, will be presented in the next section where an indirect adaptive control strategy will be discussed.

5.4 The Adaptive Pole Placement Controller

5.4.1 State Feedback Controller and State Observer

The problem of the synthesis of pole placement controllers can be discussed starting from a state space description of the plant, or by directly taking into account an input-output description based on the representation of the controlled process by transfer functions in the z-domain. For the basic theory discussed in this section any time dependence of the parameters will be neglected.

The linear single-input single-output (SISO) system is in this case defined by the state space equations

$$x(k+1) = A\ x(k) + b\ u(k)\ , \quad x(0) = x_0\ , \tag{5.90}$$

$$y(k)\ = c^T\ x(k), \tag{5.91}$$

with state vector

$$x(k) = \begin{bmatrix} x_1(k) \\ \cdot \\ \cdot \\ x_n(k) \end{bmatrix},$$

n×n system matrix

$$A\ = \begin{bmatrix} a_{11} & \cdots & a_{n1} \\ \cdot\cdot & \cdot\cdot & \cdot\cdot \\ a_{n1} & \cdots & a_{nn} \end{bmatrix},$$

input vector

$$b\ = \begin{bmatrix} b_1 \\ \cdot \\ \cdot \\ b_n \end{bmatrix},$$

output vector

$$\mathbf{c}^T = [c_1 \dots c_n] \,,$$

and input signal u(k) and output signal y(k). A block diagram representing Eq. (5.90) is depicted in Fig. 5.8.

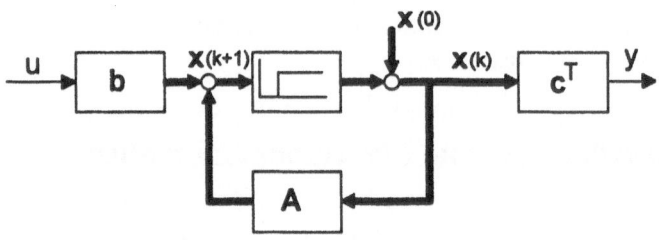

Fig. 5.8 State space description of a SISO system

In technical applications it is often the case, that the states of the plant are not directly measurable, as it happens for example for the state variable "torque" of motor speed control. To be able to implement state feedback control, an observer, which reconstructs the state signals using only measurable input and outputs signals, must be included in the control law.

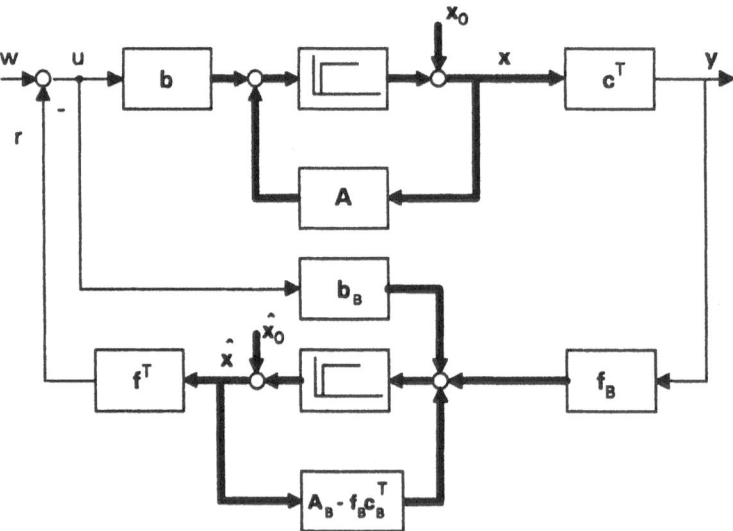

Fig. 5.9 State feedback control with state observer

A block diagram of the closed loop system with state observer is illustrated in Fig. 5.9. The intention and the purpose of this control strategy is extensively discussed in (Unbehauen, 1988). The state space approach shall here only help in a better understanding of the theory using polynomial descriptions. Here it is only important to assert that the observer estimates stationary correct state values, and that f and f_B are vectors having each n entries. As seen from the block diagram of the closed loop system, the reconstructed state vector is given by

$$\hat{x}(k+1) = (A_B - f_B\, c_B^T)\, \hat{x}(k) + b_B\, u(k) + f_B\, y(k) \tag{5.92}$$

and the feed back signal is described by

$$r(k) = f^T\, \hat{x}(k). \tag{5.93}$$

Application of the z-transformion to Eq. (5.92) leads to

$$z\, \hat{X}(z) = (A_B - f_B\, c_B^T)\, \hat{X}(k) + b_B\, U(z) + f_B\, Y(z) \tag{5.94}$$

which can be rewritten as

$$(z\, I - A_B + f_B\, c_B^T)\, \hat{X}(z) = b_B\, U(z) + f_B\, Y(z). \tag{5.95}$$

Application of the z-transformation to Eq. (5.93) results, combined with Eq. (5.95), in the frequency domain feedback law

$$R(z) = f^T\, (z\, I - A_B + f_B\, c_B^T)^{-1}\, (b_B\, U(z) + f_B\, Y(z)), \tag{5.96}$$

which can be rewritten as

$$R(z) = \frac{K(z)}{Q(z)} U(z) + \frac{H(z)}{Q(z)} Y(z) \tag{5.97}$$

with

$$\frac{K(z)}{Q(z)} = f^T\, (z\, I - A_B + f_B\, c_B^T)^{-1}\, b_B \tag{5.98}$$

and

$$\frac{H(z)}{Q(z)} = f^T\, (z\, I - A_B + f_B\, c_B^T)^{-1}\, f_B. \tag{5.99}$$

The denominator polynomial $Q(z)$ can be calculated from the determinant expression

$$Q(z) = |\, z\, I - A_B + f_B\, c_B^T\, |. \tag{5.100}$$

The roots of the polynomial $Q(z)$ are poles of the observer, this is the reason why in literature (Åström and Wittenmark, 1984) $Q(z)$ is often denoted as "observer polynomial". In Fig. 5.11 the equivalent control loop according to Eq. (5.97) is shown in polynomial description.

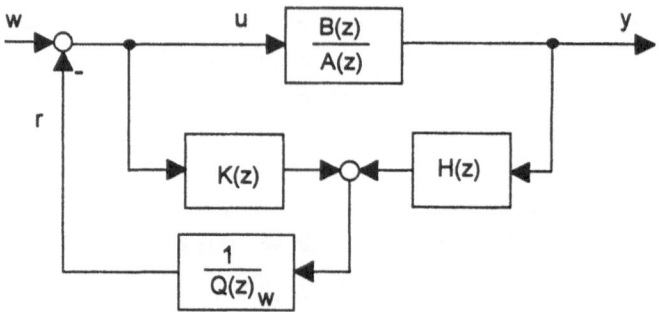

Fig 5.10 Control structure with observer and state feedback

In that case the plant is represented by the transfer function

$$\frac{Y(z)}{U(z)} = \frac{B(z)}{A(z)} \; . \tag{5.101}$$

This makes up the connection between the control law in state space description and polynomial representation. In the following solely the polynomial representation will be used, on which the synthesis of the controller will be based. The feedback structure shown in Fig. 5.11 is not well suited for the implementation on a digital computer. The algebraic loop, introduced by the feedback of the manipulated signal u, will be broken by some block manipulations according to Fig 5.11.

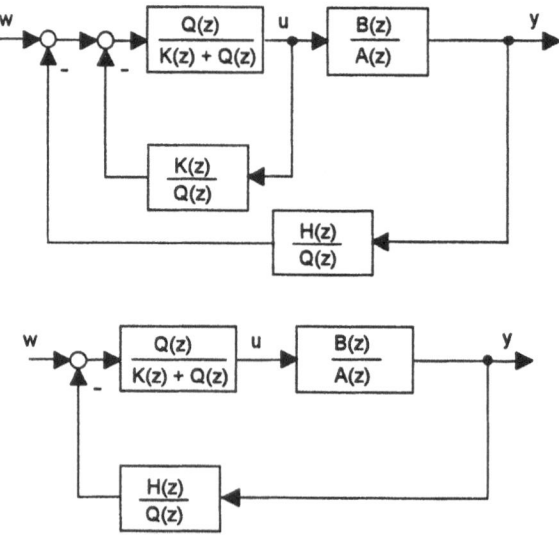

Fig 5.11 Transformation of the control structure

The elimination of the observer polynomial Q(z) from the feedback path leads to a further simplification of the structure. With

$$L(z) = K(z) + Q(z) \tag{5.102}$$

the final structure of the control loop results, as shown in Fig 5.12.

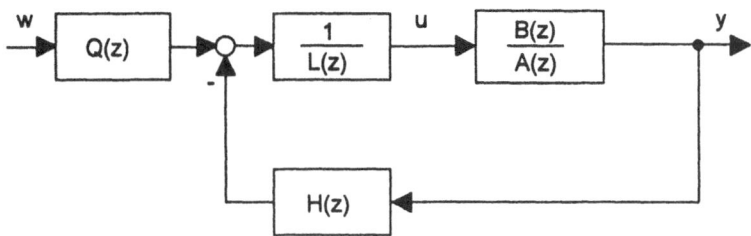

Fig 5.12 Final control structure for pole placement

The degrees of the polynomials resulting from Eqs. (5.90), (5.98)-(5.100) and (5.102) are

$$\deg\{A(z)\} = \deg\{Q(z)\} = \deg\{L(z)\} = n \ ,$$

$$\deg\{H(z)\} = n\text{-}1 \ .$$

The degree of the observer polynomial Q(z) can be reduced to n-1. The reason for this simplification is, that a reduced order observer, which estimates n-1 states, is sufficient, because the output signal is by definition measurable. Thus the degrees

$$\deg\{Q(z)\} = \deg\{L(z)\} = \deg\{H(z)\} = n\text{-}1 \tag{5.103}$$

follow for the controller polynomials. In the representation of the polynomials in z^{-1} the plant is described by

$$\frac{Y(z)}{U(z)} = \frac{B(z)}{A(z)} = \frac{b_1 z^{-1} + b_2 z^{-2} + \ldots + b_n z^{-n}}{1 + a_1 z^{-1} + \ldots + a_n z^{-n}} \ . \tag{5.104}$$

Here the coefficient b_0 is set to zero, because technical systems do not normally contain any direct feed through term. The synthesis of the observer polynomial Q(z), of the feedback polynomial H(z) and the controller polynomial L(z) will be discussed in the next sections.

5.4.2 The Linear Control Law

The controller design leads to a dynamic behaviour of the closed loop system, as chosen by the designer. Generally a transfer function of the form

$$G_w(z) = \frac{Y(z)}{W(z)} = \frac{B_w(z)}{A_w(z)} \tag{5.105}$$

can be specified for set point behaviour. With a controller according to Fig 5.12 yields a set point transfer function

$$\frac{Y(z)}{W(z)} = \frac{B(z) \, Q(z)}{A(z) \, L(z) + B(z) \, H(z)} . \tag{5.106}$$

Due to the separation principle, the observer has no influence on the setpoint transfer function of the closed loop system. This means that for the polynomial representation, the observer polynomial $Q(z)$ must be a factor of the denominator polynomial of the transfer function Eq. (5.106). Thus for synthesis purposes follows the Diophantine equation

$$A(z) \, L(z) + B(z) \, H(z) = P(z) \, Q(z) . \tag{5.107}$$

From a theoretical point of view, the parameters of the observer polynomial are freely selectable, as long as $Q(z)$ remains a stable polynomial (the practical choice will be discussed later in this chapter). The theoretical aspects of the solution of Diophantine equations are described in (Kucera, 1979). This equation has an infinite number of solutions, here the minimal degree solution is chosen. The resulting degrees with respect to the type of observer are shown in Table 5.1.

Table 5.1 Polynomial degrees for different types of observers

	reduced order observer	full order observer
n_q	n-1	n
n_a, n_b	n	n
n_p	n	n
n_h	n-1	n-1
n_l	n-1	n

For the closed loop system given by Eqs. (5.106) and (5.107) the transfer function

$$G_w(z) = \frac{Y(z)}{W(z)} = \frac{B(z)}{P(z)} \tag{5.108}$$

results. Obviously, the loop gain for setpoint behaviour in the general case is different from one. To guarantee that the stationary value of the output signal exactly follows the reference signal, a gain factor

$$K = \frac{P(1)}{B(1)} \tag{5.109}$$

will be introduced according to Fig. 5.13. K is the inverse gain of the transfer function given by Eq. (5.108).

$$G_w(z) = \frac{Y(z)}{W(z)} = K \frac{B(z)}{P(z)} \qquad (5.110)$$

is thus the final transfer function of the closed loop system.

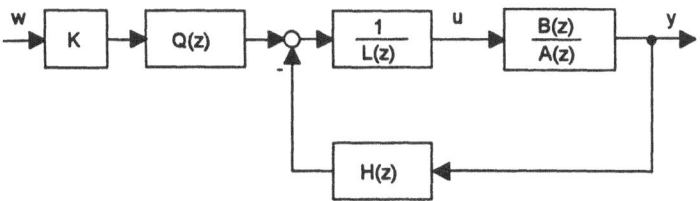

Fig 5.13 Control structure with compensation of the loop gain

The synthesis of a pole placement controller is thus done in the following steps:

- choice of closed loop poles

- choice of structure and poles of the observer Q(z)

- solution of the Diophantine equation (5.117)

5.4.2.1 A Linear Quadratic Optimal Control Strategy

The linear quadratic (LQ) regulator represents an optimal solution to the pole placement problem, which minimizes the performance criterion

$$I = \sum_{k=1}^{\infty} \{ [w(k)-y(k)]^2 + r\, u^2(k) \} . \qquad (5.111)$$

The criterion is a function of the control error

$$e(k) = w(k) - y(k) \qquad (5.112)$$

and the manipulated signal u(k), which is weighted by a positive constant factor r. This weighting enables one to influence the controller with respect to the limits of the manipulated signal, which is of great importance in real technical systems. In the state space description the minimization of the criterion function is performed by solving the algebraic matrix Riccati equation (Åström and Wittenmark, 1984). The solution of this equation leads to a linear system of equations for the coefficients of the state feedback vector f. In the SISO case the solution of the Riccati equation is equivalent to the problem (Åström and Wittenmark, 1984) of finding the stable roots of

$$I(z) = r\, A(z)\, A(z^{-1}) + B(z)\, B(z^{-1}) . \qquad (5.113)$$

This means, the n optimal poles of the closed loop system may be calculated using spectral factorization of the polynomial

$$I(z) = \Gamma \, P(z) \, P(z^{-1}) \, . \tag{5.114}$$

The 2n-degree polynomial $I(z)$ is symmetric due to the right hand side of Eq. (5.113), such that a stable spectral factor always exists. Γ is a real constant factor, which has to be introduced due to the normalization of the leading coefficient of the characteristic polynomial $P(z)$. Thus $P(z)$ can be immediately used in the pole placement algorithm, without explicitly calculating the optimal characteristic poles.

5.4.2.2 A Linear Quadratic Gaussian Control Strategy

In the previous section only plants without disturbance have been considered. In practice at least stochastic disturbances, e.g. measurement noise, will be present. Thus the model of the plant has to be augmented by a stochastic disturbance $v(k)$. This non-measurable disturbance can be modelled by filtering white noise $\varepsilon(k)$ using the filter transfer function

$$G_{sz}(z) = \frac{C(z)}{A(z)} \, . \tag{5.115}$$

The leading coefficient of the polynomial $C(z)$ should be monic as it is the case for the denominator polynomial $A(z)$. According to (Åström and Wittenmark, 1984) the quadratic performance index

$$I = E[\, e^2 \, (k) + r \, u^2 \, (k) \,] \tag{5.116}$$

has to be minimized. The solution of this optimization problem again leads to Eq. (5.113), i.e. the optimal characteristic polynomial result from spectral factorization of this equation. To implement the LQG controller the numerator polynomial $C(z)$ of the disturbance filter has to be chosen as observer polynomial (Åström and Wittenmark, 1984), so that the Diophantine equation (5.107) has to substituted by

$$A(z) \, L(z) + B(z) \, H(z) = P(z) \, C(z) \, . \tag{5.117}$$

The design algorithm of the LQG controller can be executed using the following steps:

- calculation of the characteristic polynomial $P(z)$ using spectral factorization of Eq. (5.113)

- solution of the Diophantine Equation (5.117)

- calculation of the gain factor K according to Eq. (5.109) .

5.4.3 The Adaptive Control Law

Up to now the plant parameters have been supposed to be time invariant and the coefficients of the polynomials $A(z)$, $B(z)$ as well as $C(z)$ to be known. For slowly time varying or nonlinear systems, which work in a linearized zone around the operating point this will no longer be the case, if a change of the operating point occurs. The use of the pole placement controller for such types of plants needs in addition perpetual updating of the model parameters.

5.4.3.1 Parameter Estimation

A recursive estimation algorithm generally serves for identification of plant parameters. In the case of a disturbance polynomial $C(z)=1$, acting on the plant output, the recursive least squares (RLS) method is well suited. This "ideal case" of a stochastic disturbance often fails to adequately meet the real properties of technical processes. Therefore let us assume a more general disturbance model. The algorithm for parameter estimation used for adaptive pole placement control will be a recursive maximum likelihood (RML) method.

In the time domain the plant can be described by the ARMAX model

$$y(k) + \sum_{i=1}^{n} a_i\, y(k\text{-}i) = \sum_{i=1}^{n} b_i\, u(k\text{-}i) + \sum_{i=1}^{n} c_i\, \varepsilon(k\text{-}i) \tag{5.118}$$

or equivalently by

$$y(k) = m^T (k)\, p + \varepsilon(k) \tag{5.119}$$

with parameter vector

$$p = [\, a_1 \ \ldots \ a_n \ | \ b_1 \ \ldots \ b_n \ | \ c_1 \ \ldots \ c_n \ ]^T \tag{5.120}$$

and data vector

$$m^T(k) = [\, \text{-}y(k\text{-}1) \ldots \text{-}y(k\text{-}n)\, | \, u(k\text{-}1) \ldots u(k\text{-}n)\, | \, \varepsilon(k\text{-}1) \ldots \varepsilon(k\text{-}n)\,] \tag{5.121}$$

where any time dependence of the parameters has been neglected. This is allowed for sufficiently long terms of constant or slowly time varying parameters. The task of the estimation procedure is to generate an estimate of

$$\hat{p} = [\, \hat{a}_1 \ \ldots \ \hat{a}_n \ | \ \hat{b}_1 \ \ldots \ \hat{b}_n \ | \ \hat{c}_1 \ \ldots \hat{c}_n \ ]^T . \tag{5.122}$$

For that purpose a special version of the more general recursive prediction error method (RPEM) will be proposed. The construction of this estimation procedure and its discussion can be taken from (Ljung and Söderström, 1983). Here only the pure algorithm in our notation will be presented as it has been implemented in the adaptive pole placement controller.

A recursive Maximum-Likelihood algorithm (RML)

1. Calculation of the prediction error

$$\hat{\varepsilon}(k) = y(k) - \hat{y}(k) \tag{5.123a}$$

2. Update of the covariance matrix

$$\mathbf{P}(k) = \mathbf{P}(k-1) - \frac{\mathbf{P}(k-1)\ \phi(k)\ \phi^T(k)\ \mathbf{P}(k-1)}{1 + \phi^T(k)\ \mathbf{P}(k-1)\ \phi(k)} \tag{5.123b}$$

3. Update of the parameter vector

$$\hat{\mathbf{p}}(k) = \hat{\mathbf{p}}(k-1) + \frac{\mathbf{P}(k-1)\ \phi(k)}{1 + \phi^T(k)\ \mathbf{P}(k-1)\ \phi(k)}\ \hat{\varepsilon}(k) \tag{5.123c}$$

4. Estimation of the stochastic disturbance

$$\hat{\varepsilon}_0(k) = y(k) - \hat{\mathbf{p}}^T(k)\ \mathbf{m}(k) \tag{5.124a}$$

5. Update of the data vector

$$\mathbf{m}(k) = [\ -y(k) \ldots -y(k-n+1)\ |\ u(k) \ldots u(k-n+1)\ |\ \hat{\varepsilon}_0(k-1) \ldots \hat{\varepsilon}_0(k-n)\ ]^T \tag{5.124b}$$

6. Prediction of the outputs signal

$$\hat{y}(k+1) = \hat{\mathbf{p}}^T(k)\ \mathbf{m}(k+1) \tag{5.124c}$$

7. Filtering of signals using the disturbance polynomial C(z)

$$y'(k) = y(k) - \hat{c}_1\ y'(k-1) - \ldots - \hat{c}_n\ y'(k-n) \tag{5.125a}$$

$$u'(k) = u(k) - \hat{c}_1\ u'(k-1) - \ldots - \hat{c}_n\ u'(k-n) \tag{5.125b}$$

$$\varepsilon'(k) = \hat{\varepsilon}_0(k) - \hat{c}_1\ \varepsilon'(k-1) - \ldots - \hat{c}_n\ \varepsilon'(k-n) \tag{5.125c}$$

8. Update of the gradient vector

$$\phi(k) = [\ -y'(k) \ldots -y'(k-n+1)\ |\ u'(k) \ldots u'(k-n+1)\ |\ \varepsilon'(k-1) \ldots \varepsilon'(k-n)\ ]^T \tag{5.125d}$$

Eqs. (5.123a,b,c) parallel the standard RLS method, if the gradient vector is substituted by the data vector. The stochastic disturbance is not measurable. Therefore it will be estimated using Eq. (5.124a). The resulting signal values are only exactly correct, if the parameter vector converges to "true" values and the prior n estimates of $\hat{\varepsilon}_0$ are also free of errors. Thus the data vector in Eq. (5.124b) will not generally contain correct values of the stochastic disturbance. The gradient vector is composed of filtered input and output signals, as well as of filtered values of the stochastic disturbance, which follows from the partial derivatives of the output prediction with respect to the parameters (Ljung and Söderström, 1983).

5.4.3.2 The RLS Algorithm as Basis for the RML Method

Eqs. (5.124b,c) form the update equations of the standard RLS method. From the practical point of view, this method shows bad properties. Standard RLS can not be used for time varying plants, because the covariance matrix **P** will decay in time, such that parameter variations can not be tracked (Warwick, 1988). Further, the implementation using finite word length of digital computer equipment leads to several numerical problems. Especially, the covariance matrix can lose its property of positive definiteness due to rounding errors. This will lead to instability of the estimation algorithm.

Table 5.2 Different weighted RLS methods

method	parameter / covariance update	comments
exponential weighting	$P_w(k) = \dfrac{1}{\mu} P(k)$	$0 < \mu \le 1$
error dependent weighting (Fortescue et al., 1981)	$P_w(k) = \dfrac{1}{\mu(k)} P(k)$	$0 < \mu(k) \le 1$ choice depends on prediction error
constant trace of the covariance matrix (Goodwin et al., 1984)	$P_w(k) = \dfrac{1}{\mu(k)} P(k)$	$\mu(k)$ calculated in such a way that the trace of the covariance matrix is kept constant
sigma modification (M'Saad et al., 1986)	$P_w(k) = \dfrac{1}{\mu(k)} P(k)$ $\hat{p}_w(k) = \hat{p}(k) + (1-\sigma)\,\hat{p}(k_1)$	$\mu(k)$ is calculated in such that the trace of the covariance matrix is kept constant $0 < \sigma \le 1$ ($= 1$, if sufficient excitation)
data dependent weighting (Dasgupta and Huang, 1987))	$P(k) = \dfrac{1}{1-\mu(k)} [\,P(k-1) -$ $\dfrac{\mu(k)P(k-1)\phi(k)\,\phi^T(k)P(k-1)}{1-\mu(k)+\mu(k)\,\phi^T(k)P(k-1)\phi(k)}\,]$ $\hat{p}(k) = \hat{p}(k-1) +$ $\dfrac{\mu(k)P(k-1)\phi(k)}{1-\mu(k)+\mu(k)\,\phi^T(k)P(k-1)\phi(k)}\hat{\varepsilon}(k)$	$0 < \mu(k) \le 1$ choice done with respect to measured signals and prediction error
matrix regularization (Ortega et al., 1985)	$P_W(k) = (1-\dfrac{\mu_0}{\mu_1}\,P(k) + \mu_0\,I)$	μ_0 and μ_1 are minimum and maximum eigenvalue of the **P** matrix

The decay of the covariance matrix may be suppressed by introducing a weighting or "forgetting factor". The numerical problems can be coped with using the UDU factorization according to Bierman (1977). In that case the covariance matrix will be factored as

$$\mathbf{P}(k) = \mathbf{U}(k)\,\mathbf{D}(k)\,\mathbf{U}^T(k) . \tag{5.126}$$

U is an upper triangular matrix with normalized diagonal entries.

The subroutine library for RLS algorithm in the CADACS real-time toolbox contains 12 different methods of weighting (Fabritz, 1989) which are mainly based on the methods presented in (Warwick, 1988). For the implemented quadratic optimal control strategy five optional methods according to Table 5.2 have been chosen.

5.4.3.3 Estimation of an Unstable Disturbance Polynomial C(z)

If an unstable disturbance polynomial is estimated, filtering according to Eq. (5.125a,b,c) is not allowed. In this case the stable spectral factor of

$$\hat{C}(z)\,\hat{C}(z^{-1}) \;=\; f_n\,z^n + ... + f_n\,z^n \,+\, f_0 \,+\, f_1\,z^{-1} + ... + f_n\,z^{-n} \tag{5.127}$$

has to be calculated. The polynomial $C^*(z)$, formed from the stable factors, will then be used for stable filtering. This procedure is allowed due to the spectral properties of the stochastic disturbance (Åström and Wittenmark, 1984).

5.4.3.4 Contraction of Filter Parameters

For $C(z)=1$ the RML algorithm approaches the extended least squares (ELS) method. This algorithm shows better properties during the transient phase, whereas the RML-method shows a better convergence (Werner, 1989). This means, that especially during the start up it is preferable to work with an ELS methods.

To combine both estimation schemes, the filter parameters are provided with a so called contraction factor κ so that the actual filtering is done using the polynomial

$$C_\kappa(z) = 1 + \kappa\,\hat{c}_1\,z^{-1} + \kappa^2\,\hat{c}_2\,z^{-2} + ... + \kappa^{-n}\,\hat{c}_n\,z^{-n} . \tag{5.128}$$

κ should be zero during start-up and tend to one later. In our control strategy a recursive evaluation of the contraction factor

$$\kappa(k) = \mu_k\,\kappa(k\text{-}1) + (1\text{-}\mu_k) , \quad \kappa(0) = 0 \tag{5.129}$$

is implemented, where μ_k is a positive constant marginally less than one. For $\mu_k=1$ the algorithm is of ELS type and for $\mu_k=0$ it is a pure RML method.

5.4.3.5 Normalization of the Gradient Vector

Normalization of the gradient vector increases the numerical quality of the calculations. Implemented is a normalization according to (Warwick, 1988)

$$n_F(k) = \max(1 , \| \phi(k) \|) , \tag{5.130}$$

where $\| \phi(k) \|$ is the ∞-Norm. Eqs. (5.124b,c) and their weighted versions according to Table 5.2 are evaluated using the normalized gradient vector

$$\phi_n(k) = \frac{\phi(k)}{n_F(k)} . \tag{5.131}$$

The normalization guarantees that the magnitude of all elements in this vector will be less than one. The evaluation of the prediction error, Eq. (5.124a), is in this case given by

$$\hat{\varepsilon}_n(k) = (y(k) - \hat{y}(k)) / n_F(k) \tag{5.132}$$

5.4.3.6 The Indirect Adaptive Controller

Two approaches exist to obtain the controller parameters. The first one is an implicit adaptive control strategy, where the controller parameters are estimated directly from measured input and output signals. But if only the controller output signal is used without special test signals this approach shows bad convergence properties (Yu et al. 1987). Therefore, an indirect adaptive method will be proposed here. For a direct - or explicit - adaptive method the plant parameters will be estimated and substituted into the polynomials of the quadratic optimal design. The characteristic polynomial of the control system then results, with respect to Eq. (5.113), from the spectral factorization of

$$\Gamma P(z) P(z^{-1}) = r \hat{A}(z)\hat{A}(z^{-1}) + \hat{B}(z)\hat{B}(z^{-1}). \tag{5.133}$$

The controller parameters of the LQ controller, Eq. (5.107), are evaluated using

$$\hat{A}(z) L(z) + \hat{B}(z) H(z) = P(z) Q(z) \tag{5.134}$$

the LQG controller, Eq. (5.117), results from

$$\hat{A}(z) L(z) + \hat{B}(z) H(z) = P(z) \hat{C}(z) . \tag{5.135}$$

The controller design is based on actual estimated values of the plant parameters neglecting any uncertainties, following the "certainty-equivalence principle".

5.4.4 Implementation Issues

5.4.4.1 Handling Dead Times of the Plant

In this section the transfer function of the plant, Eq. (5.104), will be augmented by a discrete delay term d, so that one obtains

$$\frac{Y(z)}{U(z)} = \frac{B(z)}{A(z)} z^{-d} = \frac{b_1 + b_2 z^{-1} + \ldots + b_n z^{-n}}{1 + a_1 z^{-1} + \ldots + a_n z^{-n}} z^{-d} . \qquad (5.136)$$

The dead time will be considered as known and time invariant. For the transfer function of the closed loop systems follows from Eq. (5.106)

$$\frac{Y(z)}{W(z)} = \frac{B(z)\, Q(z)\, z^{-d}}{A(z)\, L(z) + B(z)\, H(z)\, z^{-d}} . \qquad (5.137)$$

Table 5.3 Polynomial degrees for a dead time (d>0) of the plant

	reduced order observer	full order observer
n_q	n-1	n
n_a, n_b	n	n
n_p	n	n
n_h	n-1	n-1
n_l	n+d-1	n+d-1

To design a causal and realizable controller, the dead time of the closed loop must be the same as the dead time of the plant. Calculation of the controller polynomials is performed by solving the Diophantine equation

$$A(z)\, L(z) + B(z)\, H(z)\, z^{-d} = P(z)\, Q(z) , \qquad (5.138)$$

where in the case of adaptive control the estimated values of the coefficients of polynomials A(z) and B(z) will be used. If the product

$$B'(z) = B(z)\, z^{-d} \qquad (5.139)$$

is considered as polynomial of degree n+d, with leading d+1 coefficients equal to zero, it can be concluded that the evaluation of this equation is obtained using the same algorithm as in the case without dead time. The price to be paid for this approach is an increase in the degree of the controller polynomials. The resulting minimal degrees are shown in Table 5.3. A comparison with Table 5.1 shows, that in contrast to the plant without dead time, the degree of the controller polynomial is independent of the choice of the order of the observer.

In addition to take care of the dead time in the design of the controller, the discrete delay d must be included in the parameter estimation. Changes are necessary for the input signal terms and the data vector of Eq. (5.124b) turns into

$$m(k+1) = [-y(k) \dots -y(k-n+1) \mid u(k-d) \dots u(k-n-d+1) \mid \hat{\varepsilon}_0(k) \dots \hat{\varepsilon}_0(k-n+1)]^T.$$

$$(5.140)$$

The filtered input signal of Eq. (5.125b) will be calculated according to

$$u'(k) = u(k-d) - \hat{c}_1 u'(k-1) - \dots - \hat{c}_n u'(k-n) .$$

$$(5.141)$$

These changes in the estimation algorithm force the assumption of a known and fixed dead time.

5.4.4.2 Stochastic Disturbances

Stochastic disturbances have already been handled as explained above. The LQG controller shall minimize the influence of this type of disturbances on the controlled signal. According to Fig. 5.14, the disturbance transfer function results

$$\frac{Y(z)}{\varepsilon(z)} = \frac{C(z) L(z)}{A(z) L(z) + B(z) H(z)}$$

$$(5.142)$$

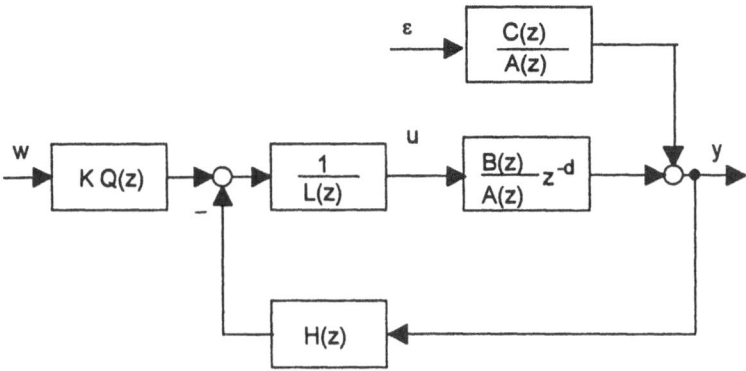

Fig. 5.14 Closed loop system with stochastic disturbance

Taking into account the Diophantine equation (5.107), the LQG disturbance transfer function can be rewritten as

$$\frac{Y(z)}{\varepsilon(z)} = \frac{C(z) L(z)}{P(z) Q(z)},$$

$$(5.143)$$

which illustrates the necessity of a stable observer polynomial $Q(z)$. For LQG control the disturbance polynomial $C(z)$ is chosen as the observer polynomial. Thus the disturbance transfer function is simplified by cancellation to

$$\frac{Y(z)}{\epsilon(z)} = \frac{L(z)}{P(z)} . \tag{5.144}$$

During adaptive control it can happen, that an unstable disturbance polynomial is estimated. Then $C(z)$ may not be used as observer polynomial, instead its stable spectral factor $C^*(z)$ has to be used.

5.4.4.3 Measurable Deterministic Disturbances

The model of the plant will now be augmented using a deterministic disturbance term. This disturbance $z(k)$ is fed to the output using the disturbance filter

$$G_{DZ}(z) = \frac{D(z)}{A(z)} z^{-d_z} = \frac{d_1 z^{-1} + ... + d_n z^{-n}}{1 + a_1 z^{-1} + ... + a_n z^{-n}} z^{-d_z} . \tag{5.145}$$

$A(z)$ is the common denominator polynomial of plant and disturbance system. The resulting system to be controlled is depicted in Fig. 5.14.

Here we will consider the case, that the disturbance transfer function $G_{DZ}(z)$ is unknown or varies in time. Taking into account that the disturbance $z(k)$ is measurable and the dead time d_z is known and time-invariant, the deterministic disturbance polynomial $D(z)$ can be estimated. In time domain the system can be described by the difference equation

$$y(k) + \sum_{i=1}^{n} a_i \, y(k-i) = \sum_{i=1}^{n} b_i \, u(k-d-i) + \sum_{i=1}^{n} d_i \, z(k-d_z-i) + \epsilon(k) + \sum_{i=1}^{n} c_i \, \epsilon(k-i)$$
$$\tag{5.146}$$

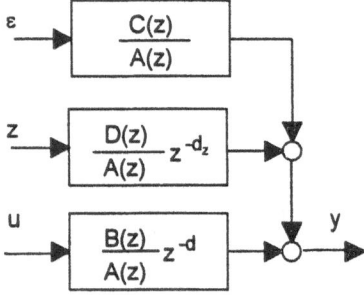

. **Fig. 5.15** Stochastic and deterministic disturbance systems

The parameter estimation algorithm has to be carried out using the data vector

$$\begin{aligned} m(k) = [\, &-y(k) \, ... \, -y(k-n+1) \, | \, u(k-d) \, ... \, u(k-n-d+1) \\ &| \, d(k-d_z) \, ... \, d(k-n-d_z+1) \, | \, \hat{\epsilon}_0(k) \, ... \, \hat{\epsilon}_0(k-n+1)]^T \end{aligned} \tag{5.147}$$

and the parameter vector

$$\hat{p} = [\, \hat{a}_1 \; \dots \; \hat{a}_n \; | \; \hat{b}_1 \; \dots \; \hat{b}_n \; | \; \hat{d}_1 \; \dots \; \hat{d}_n \; | \; \hat{c}_1 \; \dots \; \hat{c}_n \,]^T \,. \tag{5.148}$$

Additionally the gradient vector, Eq. (5.125d), has to be augmented using the last n measured disturbance values

$$d'(k) = u(k-d) - \hat{c}_1 \, d'(k-1) - \dots - \hat{c}_n \, d'(k-n) \,. \tag{5.149}$$

The introduction of the common denominator leads theoretically to common poles in the polynomials $A(z)$ and $B(z)$. In that case the Diophantine equation ((5.134) respectively (5.135)) is not solvable, because the associated linear system of equations becomes singular. Practically this problem is reduced due to uncertainties in parameter estimation as well as in the description of the real process. In all practical experiments this problem could never be observed.

Up to now only identification of the measurable disturbance, but not its compensation has been discussed. It is usually proposed to design a so called feedforward controller, which will be added to the pole placement controller. The measured disturbance will be fed to the control loop just in front of the controller polynomial $L(z)$, using the transfer function of the feedforward controller (see Fig. 5.14). For synthesis of the controller two different approaches exist. Sternad (1989) or Hunt and Grimble in (Warwick, 1988) augment the quadratic optimal approach also for the feedforward controller. This leads to the need for solving three Diophantine equations instead of one. Additionally an estimation of the z-transform of the measurable disturbance has to be carried out. Another approach chosen is the addition of a feedforward in the forward path, so that the disturbance acting on the output of the plant will be compensated (Goodwin et al., 1988).

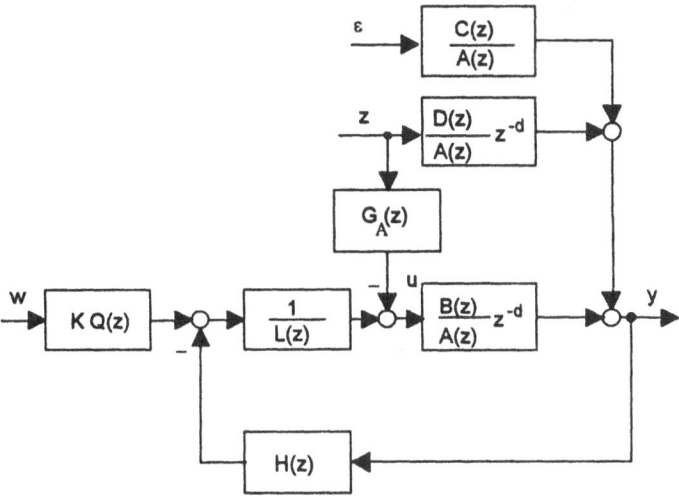

Fig. 5.16 Control structure with compensation of the deterministic disturbance

Owing to the lower complexity, the second approach has been used in this implementation. In contrary to (Goodwin et al., 1988) the feedforward is done directly to the plant input, so that the complexity will be even more reduced. The resulting control structure is shown in Fig 5.16.

The influence of the deterministic disturbance on the output signal is described by

$$\frac{Y(z)}{Z(z)} = \frac{D(z)}{A(z)} z^{-d_z} \frac{1}{1 + \frac{B(z)\,H(z)}{L(z)\,A(z)} z^{-d}} - G_A(z) \frac{B(z)z^{-d}/A(z)}{1 + \frac{B(z)\,H(z)}{L(z)\,A(z)} z^{-d}} \tag{5.150}$$

$$= \frac{L(z)}{P(z)\,Q(z)} \left(D(z)\, z^{-d_z} - G_A(z)\, B(z) z^{-d} \right).$$

The requirement

$$\frac{Y(z)}{Z(z)} \overset{!}{=} 0$$

leads to the transfer function

$$G_A(z) = \frac{D(z)}{B(z)} z^{-(d_z - d)} \tag{5.151}$$

for the feedforward term, which compensates the deterministic disturbance. For adaptive control according to the "certainty equivalence principle", the measured disturbance is fed forward to the manipulated signal using the transfer function

$$\frac{\hat{D}(z)}{\hat{B}(z)} z^{-(d_z - d)} = \frac{\hat{d}_1 + \hat{d}_2 z^{-1} + \dots + \hat{d}_n z^{-(n-1)}}{\hat{b}_1 + \hat{b}_2 z^{-1} + \dots + \hat{b}_n z^{-(n-1)}} z^{-(d_z - d)}. \tag{5.152}$$

This feedforward structure is only realizable if B(z) is minimum phase. Otherwise the controlled system would be unstable. Additionally the dead time of the disturbance system must be greater than or equal to the plant dead time

$$d_z \geq d , \tag{5.153}$$

because otherwise the feedforward would become non-causal. If these conditions are not fulfilled the dynamic feedforward has to be substituted by a static one, so that the disturbance will be at least compensated in steady state. Therefore the transfer function Eq. (5.152) will be substituted by

$$K = \frac{\hat{D}(1)}{\hat{B}(1)} . \tag{5.154}$$

5.4.4.4 Non Measurable Deterministic Disturbances

Non measurable deterministic disturbances add another problem to technical systems. The controlled signal is often influenced by constant offsets, or non localizable sinusoidal disturbances acting on the plant. Without loss of generality the non measurable disturbance shall be added to the output of the plant according to Fig. 5.17.

It will be supposed that the disturbance s is generated using a δ-impulse by the filtering

$$S(z) = \frac{E(z)}{F(z)} \, \delta(z) . \tag{5.155}$$

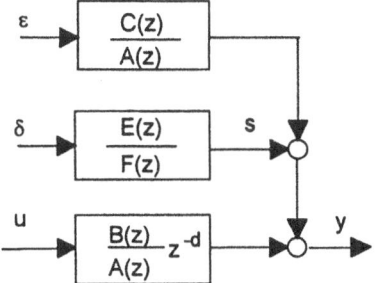

Fig. 5.17 System with stochastic and non measurable deterministic disturbances

Using the z-transform of the δ-impulse

$$\delta(z) = \delta_0 \tag{5.156}$$

follows

$$S(z) = \frac{E(z)}{F(z)} \, \delta_0 . \tag{5.157}$$

Thus the non-measurable disturbance can be represented by the disturbance filter z-transform weighted by δ_0. So the measurable output signal of the plant is given by

$$Y(z) = \frac{B(z)}{A(z)} z^{-d} U(z) + \frac{C(z)}{A(z)} \varepsilon(z) + \frac{E(z)}{F(z)} \delta(z) . \tag{5.158}$$

Multiplication by $A(z)F(z)$ on both sides leads to

$$A(z)F(z) \, Y(z) = B(z)F(z) \, z^{-d} U(z) + C(z)F(z) \, \varepsilon(z) + E(z)A(z) \, \delta(z) \tag{5.159}$$

which can be rewritten as

$$\tilde{A}(z) \, Y(z) = \tilde{B}(z) \, z^{-d} U(z) + \tilde{C}(z) \, z^{-d} \varepsilon(z) + A_E(z) \, \delta(z) , \tag{5.160}$$

if polynomials multiplied by $F(z)$ are denoted by the tilde "~". The influence of the weighting of the δ-impulses can be neglected, because the time domain examination of

$$\sum_{i=0}^{n+n_e} a_E\, \delta(k\text{-}i)$$

shows with $\delta(0)=\delta_0$ and $\delta(k)=0$, for $k \neq 0$, that this term already vanishes $n+n_e$ steps after the start of the disturbance sequence. So the system may be described by

$$\tilde{A}(z)\, Y(z) \;=\; \tilde{B}(z)\, z^{\text{-}d}\, U(z) \;+\; \tilde{C}(z)\; \delta(z)\,. \qquad\qquad (5.161)$$

The denominator polynomial of the z-transformed disturbance influences all estimated polynomials of the plant. Fernandes et al. (1987) use this fact to estimate the frequency of a non measurable periodic disturbance.

If an estimation procedure for $n+n_f$ parameters is used, all estimated polynomials must contain n_f equal poles. By factorization of all polynomials and by comparing their roots with the roots of the denominator polynomial, $F(z)$ can be evaluated.

The z-transforms of typical non measurable disturbances, constant output offset and sinusoidals, have marginally stable poles on the unit circle of the z-plane. If the controller design is performed by Eqs. (5.133) and (5.134) respectively (5.135), these marginally stable poles will not be compensated. Spectral factorization of a marginally stable pole is not possible and common poles in $A(z)$, $B(z)$ and $P(z)$ lead to singularity of the Diophantine equation.

Instead, the compensation of the dynamics of the disturbance has to be incorporated in the controller. That means that the controller polynomial $L(z)$ will be augmented by the factor $F(z)$. The Diophantine equation for this case is

$$A(z)\, L(z)\, F(z) + B(z)\, H(z)\, z^{\text{-}d} \;= P(z)\, Q(z)\,. \qquad\qquad (5.162)$$

This idea is called the "internal model principle" (Åström, 1988). Fig. 5.18 shows the resulting control structure. The motivation for this approach follows from the disturbance transfer function

$$\frac{Y(z)}{S(z)} = \frac{A(z)\, L(z)\, F(z)}{P(z)\, Q(z)} \qquad\qquad (5.163)$$

when regarding the stationary case

$$\lim_{k\to\infty} y(k,s) \;=\; \lim_{z\to 1}\, (z\text{-}1)\, \frac{A(z)\, L(z)\, F(z)\; E(z)}{P(z)\, Q(z)\; F(z)} = 0\,, \qquad\qquad (5.164)$$

because for a correct design neither $P(z)$ nor $Q(z)$ contains a marginally stable pole at $z=1$. The evaluation of the optimal characteristic polynomial has to be done with respect to the auxiliary manipulated signal (Åström and Wittenmark, 1984)

$$U_H(z) = F(z) U(z),\qquad (5.165)$$

i.e. instead of Eq. (5.116)

$$I = E[e^2(k) + r u_H^2]\qquad (5.166)$$

will be minimized. The characteristic poles are, equivalent to Eqs. (5.113) and (5.114), given by the spectral factorization

$$\Gamma P(z) P(z^{-1}) = r A(z) F(z) F(z^{-1}) A(z^{-1}) + B(z) B(z^{-1}).\qquad (5.167)$$

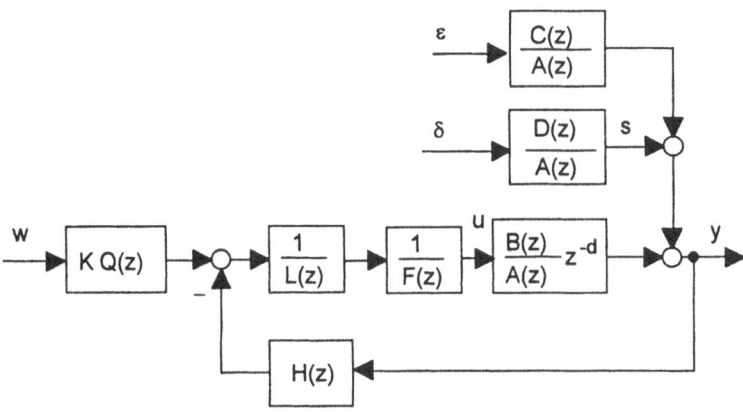

Fig. 5.18 Control structure according to the "internal model" principle

Based on this approach, the degree of the characteristic polynomial will be increased by n_f.

The main problem in the case of non-measurable deterministic disturbances is the evaluation of $F(z)$. It has been discussed above, that theoretically, due to overparametrization of the estimation problem, this polynomial will be a factor in all estimated polynomials. Simulation studies and experiments show, that this is true only in rare cases. Instead, a specific region has to be chosen, wherein poles are judged as equal and belonging to $F(z)$. This leads to great uncertainties. Additionally, the effort of the overparametrized estimation procedure and the factorization is high, so that a practical application of this approach seems to be unreasonable. To handle non-measurable disturbances it is necessary to take a fixed polynomial into account.

5.4.4.5 Special Case: Integrator and Step Disturbances

For a step disturbance one has

$$F(z) = 1 - z^{-1} \ . \tag{5.168}$$

The augmented control structure according to Fig. 5.18 is equivalent to an integral control law, its properties are summarized in Table 5.4. This is a well examined method to compensate step disturbances or offsets.

Table 5.4 Degrees of polynomials in case of integral control law

	reduced order observer	full order observer
n_q	n	n+1
n_a, n_b	n	n
n_p	n+1	n+1
n_h	n	n
n_l	max(n+d-1,n)	max(n+d-1,n+1)

For practical purposes almost always integral control laws should be used. The expenditure for inclusion of an integrator is an increase in the degree of the closed loop characteristic polynomial and in the order of the observer and controller polynomials.

5.4.4.6 Estimation of an Output Bias

The use of an integral control law increases the steady state control performance, but not the quality of parameter estimation. The convergence of the estimated plant parameters to true values is not possible in the case of biased estimation (Ljung and Söderström, 1983). To avoid this problem two methods for bias estimation have been implemented.

- Estimation as additional parameter

- Separate estimation

In the first method the parameter vector is augmented by an offset parameter y_0. The parameter estimation algorithm has to be carried out using the modified parameter vector

$$\hat{p} = [\ \hat{a}_1 \ ... \ \hat{a}_n \ | \ \hat{b}_1 \ ... \ \hat{b}_n \ | \ \hat{d}_1 \ ... \ \hat{d}_n \ | \ \hat{c}_1 \ ... \ \hat{c}_n \ | \ \hat{y}_0 \ ]^T \tag{5.169}$$

and data vector

$$
\begin{aligned}
m(k) = [\ &-y(k) \ ... \ -y(k-n+1) \ | \ u(k-d) \ ... \ u(k-n-d+1) \\
&| \ d(k-d_z) \ ... \ d(k-n-d_z+1) \ | \ \hat{\varepsilon}(k) \ ... \ \hat{\varepsilon}(k-n+1) \ | \ 1 \ ]^T \ .
\end{aligned}
\tag{5.170}
$$

The estimate is not the offset acting on the output, but the offset filtered by A(z).

For separate bias estimation Eq. (5.119) is augmented to

$$y(k) = m^T(k) \; \hat{p}(k) + \hat{\varepsilon}(k) + \hat{y}_0 = m^T(k) \; \hat{p}(k) + \hat{\varepsilon}_s(k) \; . \qquad (5.171)$$

If the offset is constant for a longer period, it can be viewed as mean value of the estimation error $\hat{\varepsilon}_S$, because the prediction error $\hat{\varepsilon}$ has stationary zero mean. Thus bias estimation is reduced to recursive evaluation of the mean value through

$$\hat{y}_0(k) \; = \; \frac{(k-1) \; \hat{y}_0(k-1) + \hat{\varepsilon}_s(k)}{k} \; = \; \hat{y}_0(k-1) + \frac{1}{k} \; (\; \hat{\varepsilon}_s(k) - \hat{y}_0(k-1)) \; . \qquad (5.172)$$

With increasing time the second term tends to zero. Therefore the factor $1/k$ must be substituted by a constant expression. Eq. (5.172) will be approximated by

$$\hat{y}_0(k) \; = \; \hat{y}_0(k-1) + (1 - \text{ß}) \; \hat{e}_0(k) \qquad (5.173)$$

with

$$\hat{e}_0(k) \; = \; y(k) - m^T(k) \; \hat{p}(k) - \hat{y}_0(k-1) \; = \; y(k) - \hat{y}(k) \; - \hat{y}_0(k-1) \; . \qquad (5.174)$$

This method has been proposed by Clarke and Gawthrop (see Warwick, 1988). ß is a positive, real constant less than one. The estimation algorithm changes too, so that Eq. (5.124a) has to be substituted by Eq. (5.174) and furthermore the estimate of the stochastic disturbance has to be evaluated, instead of using Eq. (5.124a), by

$$\hat{e}_0(k) \; = \; y(k) - m^T(k) \; \hat{p}(k) - \hat{y}_0(k-1) \; . \qquad (5.175)$$

5.4.4.7 Feed Forward of the Estimated Bias

The estimate of the offset offers a further possibility to compensate this disturbance. Based on the estimation the offset can be handled as a measurable disturbance using

$$D(z) = 1 \; , \; d_z = 0 \; .$$

The feedforward to the manipulated signal according to Eq. (5.154) is obtained from

$$u_A(k) \; = \; \frac{1}{\hat{B}(1)} \hat{y}_0 \; . \qquad (5.176)$$

Assuming the convergence of the estimate to the true offset value, the static feedforward can substitute the integral control law. The design problem for the controller is of lower order, but especially in stochastically disturbed systems the assumptions are often not fulfilled.

5.4.4.8 Limitation of the Manipulated Signal

The magnitude and the increment of the manipulated signal are naturally limited for all technical applications. The optimal pole placement design incorporates a weighting of the manipulated signal, which can be used to avoid reaching the limit of the controller output, but a too heavy weighting will often lead to an undesirably sluggish dynamic behaviour of the controlled system.

It is usual, that the manipulated signal reaches the limits in practical applications, especially because it is influenced additionally by feedforward for disturbance compensation. This leads to a nonlinear dependence between the output of the controller and the input signal of the plant.

For pure tracking of the reference signal the limiter for the manipulated signal is connected directly in front of the plant. The limited manipulated value u_B has to be recalculated with respect to the actual manipulated value, i.e. u_H will be set to the value, which would generate u_B without any limitation (without integrator one obtains $u_H = u_B$). For the disturbance regulation discussed here also the feedforward must be considered. Fig. 5.19 shows the section of interest in the control circuit, where an integrator is used to cope with non measurable step disturbances. The feedforward u_A comprises compensation of measurable disturbances and the offset compensation.

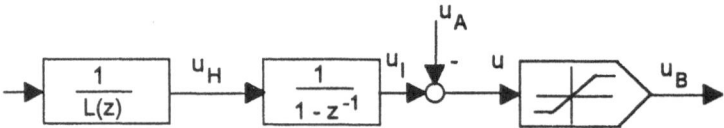

Fig 5.19 Limiting the manipulating value

It is obvious that a limitation influences the controller output as well as the feed forward. This introduces the problem, how to divide the limited value to both signals during recalculation. In the implemented version it is first tested, whether the manipulated signal without feedforward does not reach the limit. If this is the case, the limitation is only recalculated for the feedforward. Otherwise recalculation regarding only u_H is carried out and the feedforward is set to zero. If

$$| u_H(k) - u_A(k) | > u_{Max} \; ,$$

where u_{Max} is the maximum amplitude of the manipulated signal, one obtains for the evaluation of the manipulated signal without integral action

$$u_B(k) = \text{sign}(u_H(k) - u_A(k)) \, u_{Max} \tag{5.177}$$

$$u_A^* = \begin{cases} \text{sign}(u_A(k))\, u_{Max} + u_H(k) & \text{if } |u_H(k)| < u_{Max} \\ 0 & \text{else} \end{cases}$$

and

$$u_H^* = \begin{cases} u_H(k) & \text{if } |u_H(k)| < \delta u_{Max} \\ \text{sign}(u_H(k))\, u_{Max} & \text{else} \end{cases}.$$

If

$$|u_I(k) - u_A(k)| > u_{Max}$$

one obtains for the controller with integral control action

$$u_B(k) = \text{sign}(u_I(k) - u_A(k))\, u_{Max}, \tag{5.178}$$

$$u_A^* = \begin{cases} \text{sign}(u_A(k))\, u_{Max} + u_I(k) & \text{if } |u_I(k)| < u_{Max} \\ 0 & \text{else} \end{cases},$$

$$u_H^* = \begin{cases} u_H(k) & \text{if } |u_I(k)| < u_{Max} \\ \text{sign}(u_I(k))\, u_{Max} + u_B(k-1) - u_I(k-1) & \text{else} \end{cases},$$

and

$$u_I^* = \begin{cases} u_I(k) & \text{if } |u_I(k)| < u_{Max} \\ u_B(k) & \text{else} \end{cases}.$$

An analogous problem is the limitation of the increment of the manipulated signal. The magnitude of the maximum increment will be denoted as δu_{Max}. The recalculation for the control loop without integral action, necessary for

$$|u_H(k) - u_A(k) - u_B(k-1)| > \delta u_{Max}$$

with

$$\delta u_H(k) = u_H(k) - u_B(k-1)$$

is described by

$$u_B(k) = \text{sign}(u_H(k) - u_A(k))\, \delta u_{Max} + u_B(k-1), \tag{5.179}$$

$$u_A^* = \begin{cases} \text{sign}(u_A(k))\, \delta u_{Max} + u_H(k). & \text{if } \delta u_H(k)| < \delta u_{Max} \\ 0 & \text{else} \end{cases}$$

and

$$u_H^* = \begin{cases} u_H(k) & \text{if } |\delta u_H(k)| < \delta u_{Max} \\ \text{sign}(u_H(k))\, \delta u_{Max} + u_B(k-1) & \text{else} \end{cases}.$$

For the control loop with integrator respectively follows, if

$$|u_I(k) - u_A(k) - u_B(k-1)| > \delta u_{Max}$$

with

$$\delta u_H(k) = u_H(k) - u_B(k-1)$$

$$u_B(k) = \text{sign}(u_I(k) - u_A(k)) \delta u_{Max} + u_B(k-1), \tag{5.180}$$

$$u_A^* = \begin{cases} \text{sign}(u_A(k)) \delta u_{Max} + u_I(k) & \text{if} \, |\delta u_I(k)| < \delta u_{Max} \\ 0 & \text{else} \end{cases},$$

$$u_H^* = \begin{cases} u_H(k) & \text{if} \, |\delta u_I(k)| < \delta u_{Max} \\ \text{sign}(u_I(k)) \delta u_{Max} + u_B(k-1) - u_I(k-1) & \text{else} \end{cases},$$

and

$$u_I^* = \begin{cases} u_I(k) & \text{if} \, |\delta u_I(k)| < \delta u_{Max} \\ u_B(k) & \text{else} \end{cases}.$$

It is only necessary to change the feedforward value, if a dynamic feedforward is active. The symbol "*" serves to distinguish between the variables with the same name resulting from the standard algorithm.

5.4.4.9 Minimization of Time Between Measurement and Control Output

The theoretical course of the control algorithm presented above is as follwing :

- Measurement of the controlled signal and of the measurable disturbance

- Parameter estimation

- Controller design

- Calculation of the manipulated value

- Output of the manipulated value

During the evaluation of the above steps the manipulated signal $u(k)$ and the measured signals $y(k)$ and $z(k)$ are associated with the same time instants. This is obviously incorrect, because the digital computer needs a finite time for the calculations. This will lead to a computation dead time between measurement and output of the manipulated signal. To minimize this delay the parameters are estimated based on the measured values of the last sampling step. This means that estimation and controller design must be carried out prior to the measurements. The manipulated signal should also be calculated partially in advance.

Table 5.5 Evaluation of the indirect adaptive (optimal) pole placement controller

Specification of the set point
[filter set point using prefilter]
Precalculation manipulated signal
[Feedforward for compensation of] [- measurable deterministic disturbance] [- non measurable constant disturbance (estimated offset)]
Measurement of the controlled value [and deterministic disturbance]
Calculation of the manipulated value
[Feedforward for measurable deterministic disturbance]
Calculation of the actual manipulated value u_B [and the integrators]
Output of the manipulated signal
Parameter estimation: - recursive maximum-likelihood method, - extended least squares or - least-squares
Design of the quadratic optimal controller - LQ controller or - LQG controller
Evaluation of the controller performance

Therefore the calculation of the controller output without any deterministic disturbance will be discussed now. Conforming to Fig. 5.8, for the manipulated signal follows

$$L(z)\, U(z) = K\, Q(z)\, W(z) - H(z)\, Y(z) \ . \tag{5.181}$$

Respectively in time domain

$$\sum_{i=0}^{n_l} l_i\, u(k\text{-}i) \;=\; K \sum_{i=0}^{n_q} q_i\, w(k\text{-}i) \;-\; \sum_{i=0}^{n_h} h_i\, y(k\text{-}i) \ , \tag{5.182}$$

where K is the gain factor according to Eq. (5.109). In the case of LQG control $Q(z)$ represents the stabilized estimate of the disturbance polynomial $C(z)$.

The manipulated value is calculated from

$$u(k) = K \sum_{i=0}^{n_q} q_i\, w(k\text{-}i) + \sum_{i=1}^{n_l} l_i\, u(k\text{-}i) - \sum_{i=0}^{n_h} h_i\, y(k\text{-}i) .$$ (5.183)

$l_0 = 1$ in all cases, if the pole placement equation is solved using monic polynomials $A(z)$, $Q(z)$ and $P(z)$. In the difference equation (5.183) only the signal $y(k)$ is unknown before the actual measurement. Taking into account the known samples, gives

$$u(k) = u^-(k) - h_0\, y(k) .$$ (5.184)

Here

$$u^-(k) = K \sum_{i=0}^{n_q} q_i\, w(k\text{-}i) + \sum_{i=1}^{n_l} l_i\, u(k\text{-}i) - \sum_{i=1}^{n_h} h_i\, y(k\text{-}i) .$$ (5.185)

can be calculated in advance and after measurement, the last term in the sum will be added.

In an analogous way one can handle the compensation of measurable disturbances z. According to Eq. (5.152) follows for the dynamic feed forward in time-domain

$$u_A(k) = \frac{1}{\hat{b}_i} [\sum_{i=0}^{n-1} \hat{d}_{i+1}\, z(k\text{-}d_z\text{+}d\text{-}i) - \sum_{i=1}^{n-1} \hat{b}_{i+1}\, u_A(k\text{-}i)] .$$ (5.186)

If the dead time of the deterministic disturbance filter is greater than that of the plant, i.e.

$$d_z > d ,$$

then the feedforward for the manipulated signal can be calculated before the measurement has taken place. For

$$d_z = d$$

follows, that

$$u_A^-(k) = \frac{1}{\hat{b}_i} [\sum_{i=1}^{n-1} \hat{d}_{i+1}\, z(k\text{-}d_z\text{+}d\text{-}i) - \sum_{i=1}^{n-1} \hat{b}_{i+1}\, u_A(k\text{-}i)] .$$ (5.187)

can be calculated in advance. And after the measurement the feedforward may is obtained as

$$u_A(k) = u_A^-(k) + \frac{\hat{d}_1}{\hat{b}_1}\, z(k) .$$ (5.188)

For static disturbance compensation the feedforward factor in Eq. (5.154) can be evaluated prior to the measurement. Table 5.5 summarizes the evaluation of the control algorithm.

5.4.4.10 Choice of Observer Poles

For the LQG controller the observer polynomial is fixed by the stochastic disturbance polynomial $C(z)$ or its stable spectral factor $C^*(z)$. For the LQ controller neither the degree nor the parameters $Q(z)$ are assigned during its design. But in previous sections $Q(z)$ has been interpreted as observer polynomial, thus its degree will be chosen equal to the order of the plant or one less for a reduced order observer. The stability of the observer is a necessary condition. By observing the closed loop tracking and regulation behaviour follows, that the observer dynamics represented by $Q(z)$ is only important for regulation. The roots of $Q(z)$ should therefore provide a well damped asymptotic behaviour.

Two possibilities for an automatic procedure to choose the observer poles will be presented here. From state space design it is known that the observer error should decay faster than the control error. That means that the observer poles should be shifted nearer to the origin of the complex z-plane than the closed loop poles. A natural (but often bad) choice is to specify deadbeat behaviour for the observer, i.e. all poles are equal to zero. A better way is to place observer poles based on closed loop poles.

This choice of observer poles can be accomplished by setting

$$q_{Pi} = \beta_d \ p_{Pi} \quad , i=1,...,n ,$$ (5.189)

with a contraction factor

$$0 \leq \beta_d \leq 1 .$$ (5.190)

This leads to a simple connection between the coefficients of the observer polynomial and the coefficients of the characteristic polynomial according to

$$q_i = \beta_d^i \ p_i \quad , i=1,...,n .$$ (5.191)

Thus the observer polynomial can be directly derived from the parameters of the optimal closed loop polynomial. The choice of the contraction factor is up to the user.

5.5 Conclusion

This chapter considered indirect adaptive pole placement control and direct model reference adaptive control. The estimation algorithms proposed have been of the

ML-/LS-type with time-varying forgetting factor. For pole assignment robust and quadratic optimal design methodology can be used. The design also includes deterministic disturbances by introducing a general disturbance compensation.

Nearly all the items discussed as "implementation issues" for the LQ and pole placement design apply also to the model reference adaptive control scheme. Numerical precautions as factorization of the covariance matrix and on-line stability test for the estimated disturbance filter as well as procedures for integrator-windup reset practically turned out to be as important as theoretical proofs for convergence and stability of the adaptation algorithms.

5.6 References

Åström, K.J. and B. Wittenmark (1984), *'Computer controlled systems'*, Prentice Hall, Englewood Cliffs.

Åström, K. J. (1988), 'Robust and adaptive pole placement', Proc. of the American Control Conference, Vol. 3, pp. 2423-2428.

Åström, K. J. and B. Wittenmark (1980), 'Self-tuning controllers based on pole-zero placement', *IEE Proceedings*, 127.

Åström, K.J. and B. Wittenmark (1989), *'Adaptive Control'*, Addison-Wesley, Reading, Massachusetts.

Bierman, G.J. (1977), *'Factorization methods for discrete sequential estimation'*, Academic Press, New York.

Dasgupta, S. and Y.F. Huang (1987), 'Asymptotically convergent modified recursive least squares with data dependent updating and forgetting factor for systems with bounded noise', IEEE Trans. on Inform. Theory, 33, pp. 383-392.

Elliot, H. (1982), 'Direct adaptive pole placement with applications to nonminimum phase systems', *IEEE Transactions on Automatic Control*, 27, pp. 720-722.

Fabritz, N. (1989), 'Development and test of a subroutine library for weighted LS estimation algorithms', study thesis ESR-8929 (in German), Ruhr-University Bochum.

Fernandes, J., G. Goodwin and C. De Souza (1987), 'Estimation of models for systems having deterministic and random disturbances', Prepr. 10th IFAC World Congress on Automatic Control, Vol. 10, pp. 370-375.

Fortescue, F.R., L.S. Kershenbaum and B.E. Ydstie (1981), 'Implementation of self-tuning regulators with variable forgetting factor', *Automatica*, 17, pp. 831-835

Goodwin, G.C., M. Salgado and R. Middleton (1988), 'Indirect adaptive control: An integrated approach', Proc. of the American Control Conference, Vol. 3, pp. 2440-2445.

Goodwin, G. C., P.J. Ramadge and P.E. Cains (1981), 'Discrete time stochastic control', *SIAM Journ. on Control and Opt.*, 19, pp. 829-853.

Goodwin, G. C. and K. S. Sin (1981), 'Adaptive control of nonminimum phase systems', *IEEE Transactions on Automatic Control*, 26, pp. 478-483.

Goodwin, G. C. and K. S. Sin (1984), *'Adaptive Filtering Prediction and Control'*, Prentice Hall, Englewood Cliffs, N. J..

Goodwin, G.C., D.J. Hill and M. Palaniswami (1984), 'Towards an adaptive robust controller`', Proc. of the IFAC Conference on Identification and System Parameter Estimation, York, pp. 997-1002.

Hahn, V. and H. Unbehauen (1982), 'Direct adaptive control schemes for nonminimum phase systems', Prepr. of the IEEE Conference on Applications of Adaptive and Multivariable Control, Hull, pp. 170-175.

Hahn, V. (1983), 'Direct adaptive control schemes for discrete-time control of multivariable systems', (in German), Dr.-Ing. Thesis, Ruhr-University Bochum, Germany.

Ionescu, T. and R.V. Monopoli, 'Discrete model reference adaptive control with an augmented error signal', *Automatica*, 13, pp. 507-517.

Kucera, V. (1979), *'Discrete Linear Control. The Polynomial Equation Approach'*, John Wiley, Chichester.

Li, M. and M. Bayoumi (1989), 'Adaptive decoupling control of MIMO systems', *International Journal of Adaptive Control and Signal Processing*, 3, pp. 375-393.

Ljung, L. and T. Söderström (1983), *'Theory and practice of recursive identification'*, MIT Press, Cambridge, MA.

Landau, Y.D. (1979), *'Adaptive Control: The Model Reference Approach'*, Marcel Dekker Inc., New York/Basel.

Lozano, R. and I.D. Landau (1981), 'Redesign of explicit and implicit discrete time model reference adaptive control schemes', *Int. J. Control*, 33, pp. 247-268.

Lozano Leal, R. and G. C. Goodwin (1985), 'A globally convergent adaptive pole placement algorithm without a persistency of excitation requirement', *IEEE Transactions on Automatic Control*, 30, pp. 795-798.

Lozano Leal, R. and I. D. Landau (1982), 'Quasi-direct adaptive control for nonminimum phase systems', *Journal of Dynamic Systems, Measurement, and Control*, pp. 104-112.

M'Saad, M., M. Duque and I.D. Landau (1986), 'Practical implicatrions of recent results in robustness of adaptive control schemes', Proc. of the IEEE Conf. on Decision and Control, Athens, pp. 477-482.

Monopoli, R.V. (1974), 'Model reference adaptive control with an augmented error signal', *IEEE Trans. on Aut. Contr.*, 19, pp. 474-484.

Ortega, R.L., L. Praly and I.D. Landau (1985), 'Robustness of discrete-time adaptive controllers', IEEE Trans. on Autom. Contr., 30, pp. 1179-1187.

Peterson, B.B. and K.S. Narendra (1982), "Bounded error adaptive control', *IEEE Trans. on Aut. Contr.*, 27, pp. 1161-1168.

Rohrs, C.E. and K. Shortelle (1984), 'Conditioning a plant for frequency selective adaptive control with improved robustness', Proc. of the American Control Conference, San Diego, pp. 1578-1583.

Stephan, R.M, V. Hahn, J. Dastych and H. Unbehauen (1991), 'Adaptive and Robust Cascade Schemes for Thyristor Driven DC-motor Speed Control', *Automatica*, Vol. 27, pp. 449-461.

Sternad, M. (1989), 'The use of disturbance measurement feedforward in LQG self-tuners', Prepr. IFAC Symposium in Control and Signal Processing, Vol. 2, pp. 353-358.

Unbehauen, H. und P. Wiemer (1985), 'Application of multivariable adaptive control schemes to distillation columns', Prepr. 4th Yale Workshop on Application of Adaptive System Theory, Yale, pp. 23-29.

Unbehauen, H. (1988), '*Regelungstechnik II. Zustandsregelung, digitale und nichtlineare Regelsysteme*', Vieweg Verlag, Braunschweig.

Unbehauen, H. (1989), 'Entwurf und Realisierung neuer adaptiver Regler nach dem Modellvergleichsverfahren', *Automatisierungstechnik*, 37, pp- 249-257.

Unbehauen, H. and U. Keuchel (1992), 'Model reference adaptive control applied to electrical machines', *International Journal of Adaptive Control and Signal Processing*, 6, pp. 95-109.

VDI/VDE GMA 3685 (1990), 'Adaptive Controllers, Sheet 1: Properties of adaptive Control Instruments', Beuth Verlag, Berlin.

VDI/VDE GMA 3685 (1991), 'Adaptive Controllers, Sheet 2: Comments and examples', Beuth Verlag, Berlin.

Vostry, Z. (1979), 'New algorithm for polynomial spectral factorization', *Kybernetika*, 8, pp. 448-470.

Warwick, K. (ed.) (1988), '*Implementation of self-tuning controllers*', Peter Peregrinus, London.

Werner, H. (1989), 'Development of robust quadratic optimal self-tuning controllers', Dipl-Ing. thesis ESR 8911, Ruhr-Universität Bochum.

Wiemer, P. and H. Unbehauen (1990), 'On the stability of decentralized discrete adaptive control systems', *Int. J. of Adaptive Control and Signal Processing*, Vol. 4, pp. 415-437.

Wiemer, P., G. Olejua Torres and H. Unbehauen (1988), 'A robust adaptive controller for systems with arbitrary zeros', Prepr. Int. IEE Conf. CONTROL 88, Oxford, pp. 598-603.

Wiemer, P. (1988), 'Entwurf dezentraler adaptiver Regelsysteme nach der Sektor-stabilitätstheorie', Dr.-Ing. thesis, Ruhr-Universität Bochum.

Wittenmark, B., R. Middleton and G.C. Goodwin (1987), 'Adaptive decoupling of multivariable systems', *Int. Journal of Control*, 46, pp. 1993-2009.

Yu, J., U. Keuchel and H. Unbehauen (1987), 'Adaptive pole placement control of a dead-time process with an integrator and of a double integral plant', Proc. 5th Yale Workshop on Application of Adaptive System Theory, Yale, pp. 90-95.

Zhang, W.-Y., Z.-J. Wang, H.-Q. Yang and X.-H. Zhang (1989), 'Decoupling-adaptive control of a class of multivariable systems', Prepr. IFAC Symposium on Control and Signal Processing, Vol. 1, pp. 171-174.

CHAPTER 6
CONTROLLERS SYNTHESIS AND REALIZATION

6.1 Introduction

Different control structures were compared experimentally: a commercially available , a completely adaptive digital scheme and a completely adaptive hybrid scheme. In the latter one the speed controller was designed based on to techniques: the model reference principle or according to a linear quadratic optimal control law. In this chapter these structures will be explained and the controller parameters for the digital and analog current control loops and for the analog speed controller will be established.

6.2 The Commercially Available Analog Controller

The commercially available analog controller is a product of the Siemens company. Similar units are sold by several competitors, as for example AEG, Asea-Brown-Boveri, Mannesmann-Demag, WEG and many others. It consists of a dual-mode adaptive inner current loop, cascaded with a PI-speed regulator loop. The current controller switches from a PI-structure, when current flows, to an I-structure, when there is no current flow. Fig. 6.1 presents a complete diagram of the system including the plant. The analog controller parameters were optimized using the well known criteria of Kessler (Kessler, 1955; Umland et al., 1990). The application of these criteria for the analog DC-motor speed control problem which is a standard procedure in Germay, see for example Bystron (1979) or Buxbaum and Schierau (1980).

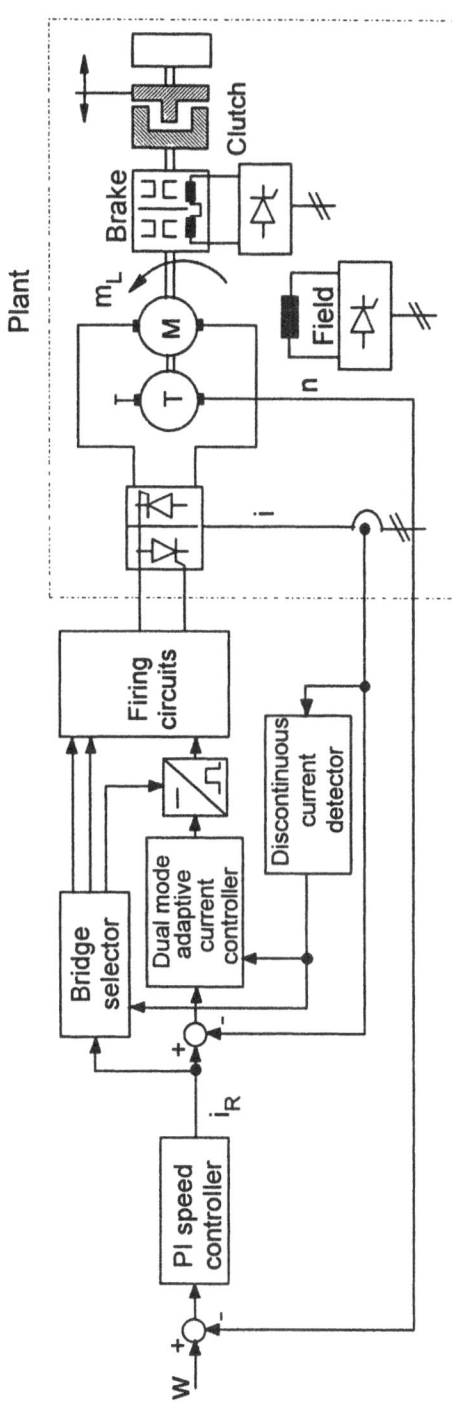

Fig. 6.1 The commercially available analog controller with inner dual-mode adaptive current controller

6.2.1 The Analog Dual-Mode Adaptive Current Controller

The inner current controller is dual-mode adaptive: in continuous current mode it is a PI-controller, in discontinuous current mode it is an I-controller with variable gain (Buxbaum, 1969). The complete plant model has already been presented in Fig. 3.2. Regarding the motor back voltage e* as a disturbance, the inner current loop has the form presented in Fig. 6.2.

If the rectifier variable dead time ($0 < T_t < 10$ ms) is represented by a statistical mean-value T_{to} of 5 ms (Lagasse and Prajoux, 1974) and applying Kessler's "Betrags-Optimum"-method (Kessler, 1955; Umland and Safuddin, 1990), the optimal controller parameters for continuous current are determined as

$$T_2 = T_A = 30 \text{ ms} , \tag{6.1}$$

$$K_2 = \frac{T_A}{2K_t K_A K_i (T_{to} + T_{ii})} = 0.41 , \tag{6.2}$$

$$T_i = T_{ii} = 2.35 \text{ ms} . \tag{6.3}$$

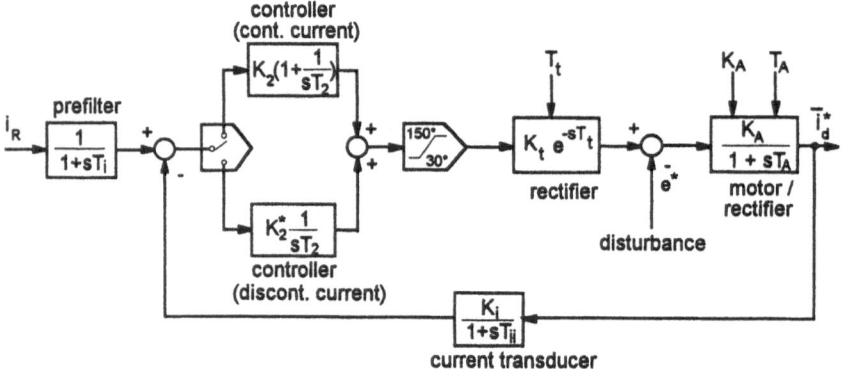

Fig. 6.2 The analog dual-mode adaptive current control loop

For the discontinuous current operating mode, the I-controller must be such that the linearized open-loop transfer function remains equal to the open-loop transfer function in continuous current operating mode. Then

$$K_2^* \frac{1}{sT_2} K_{A(\text{disc.})} = K_2 (1 + \frac{1}{sT_2}) \frac{K_{A(\text{cont.})}}{1 + sT_A} , \tag{6.4}$$

where $0.3 < K_{A(\text{disc.})} < 2$ represents the nonlinear gain K_A in discontinuous current mode, and $K_{A(\text{cont.})} = 2$ represents the gain K_A in continuous current mode. Using Eq.(6.1), it follows

$$K_2^* = K_2 \frac{K_{A(cont.)}}{K_{A(disc.)}} .$$ (6.5)

The Siemens' analog controller SIMOREG allows the adjustment of the variable gain K_2^* with a potentiometer. The parameters K_2, T_2 and T_i are fixed with resistors and capacitors.

It is important to mention that the inner current loop performs more than a simple control task. Actually, it is also responsible for the functions necessary during a change in the direction of the load current, as follows:

- detection of a change in the sign of the reference current i_R, anticipating a change in the direction of the load current,

- triggering the thyristors at $150°$ to bring the load current as rapidly as possible to zero,

- removal of the firing pulses from the conducting thyristors bridge when the current is zero,

- after a delay time to guarantee the turn-off of all thyristors, application of the firing pulses to the other bridge, beginning with firing pulses at $150°$.

These functions belong to the "bridge selector" block shown in Fig. 6.1.

6.2.2 The Analog PI Speed Controller

The outer speed loop is represented in Fig. 6.3. The dual-mode adaptive inner current loop is substituted by an equivalent first order system with time constant

$$T_e = 2 (T_{to} + T_{ii})$$ (6.6)

(Bystron, 1979; Buxbaum and Schierau, 1980). For the speed controller synthesis, nominal field excitation $\psi^* = 1$ and negligible friction coefficient $B \approx 0$ will be considered. Using Kessler's "Symmetrisches-Optimum"-method (Kessler, 1955; Umland and Safuddin, 1990), the speed controller parameters are obtained as

$$T_1 = 4 (T_e + T_{dd}) = 102.8 \text{ ms} .$$ (6.7)

$$K_1 = \frac{T_H}{2\frac{K_d}{K_i}(T_e + T_{dd})} = 7.57 ,$$ (6.8)

and for the pre-filter

$$T_d = T_1 = 102.8 \text{ ms}.$$ (6.9)

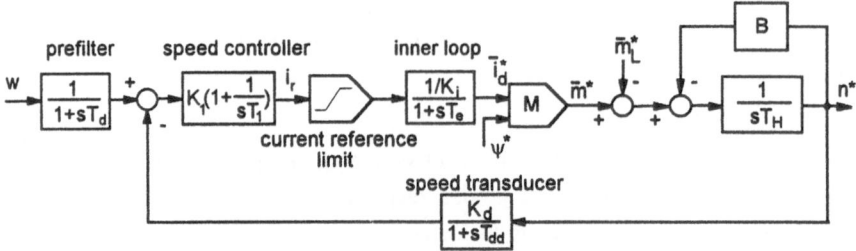

Fig. 6.3 Equivalent block diagram for the outer speed loop

These parameters were adjusted in the commercially available SIMOREG unit with resistors and capacitors. A current reference limiter guarantees that the thyristor maximal current will not be reached.

6.3 The Completely Adaptive Digital Scheme

The digital adaptive scheme is shown in Fig. 6.4. It consists of a dual-mode adaptive inner current loop and of a model reference adaptive outer speed loop (Stepan et al., 1988; Stephan, 1991).

6.3.1 The Digital Dual-Mode Adaptive Current Controller

Like the analog dual-mode adaptive current controller, the digital current controller switches from a PI-structure, in continuous current, to an I-structure, in discontinuous current. The inner digital current loop is presented in Fig. 6.5. The current measurement is made with the interface circuit described in section 3.2.2 and is modelled as a dead time of 10 ms.

For the continuous current mode, the PI-controller parameters are determined by the discrete "Betrags-Optimum"-method (Depping and Voits, 1985). The digital controller is then given by

$$K_R(z) = K_2 \left(1 + \frac{\Delta T}{T_2} \frac{1}{1-z^{-1}} \right) , \qquad (6.10)$$

where ΔT is the sampling time and K_2 and T_2 are determined by

$$T_2 = T_A - \frac{\Delta T}{2} , \qquad (6.11)$$

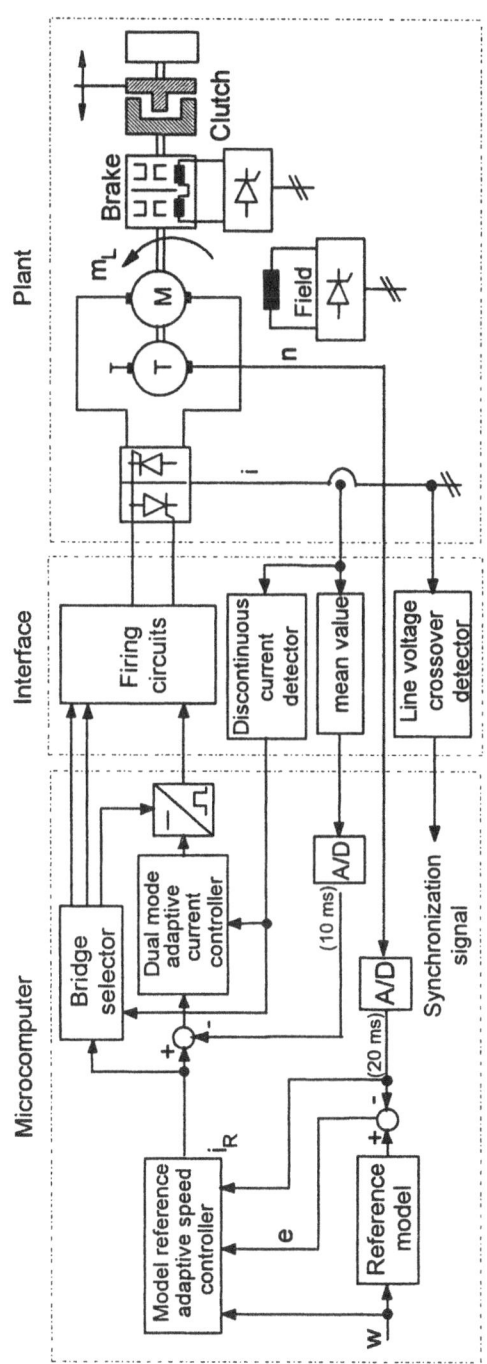

Fig. 6.4 The completely adaptive digital scheme with model reference speed controller

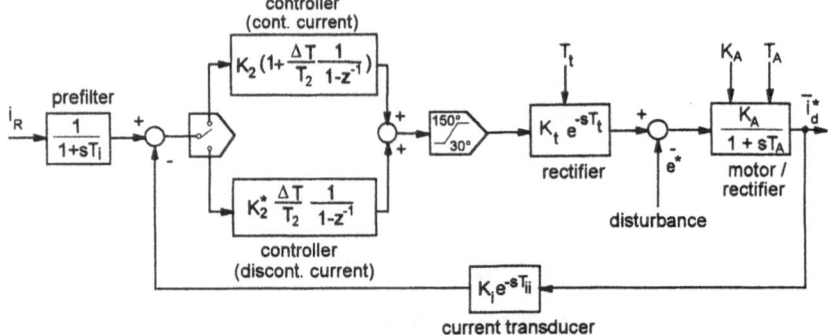

Fig. 6.5 The digital dual-mode adaptive current loop

$$K_2 = \frac{T_A - \dfrac{\Delta T}{2}}{2K_t K_A K_i (T_{t0} + T_{ii} + \dfrac{\Delta T}{2})} , \qquad (6.12)$$

where T_{t0}=5 ms is the statistical mean-value representation of the rectifier variable dead time T_t. Note that for ΔT=0 Eqs. (6.11) and (6.12) are respectively equal to Eqs.(6.1) and (6.2), thus showing the consistency between the "Betrags-Optimum" and the discrete " "Betrags-Optimum" methods. For ΔT=10 ms, Eqs. (6.11) and (6.12) give

$$T_2 = 25 \text{ ms} , \qquad (6.13)$$

$$K_2 = 0.142. \qquad (6.14)$$

The I-controller for the discontinuous current operating mode is given by

$$K_R(z) = K_2^* \frac{\Delta T}{T_2} \frac{1}{1-z^{-1}} . \qquad (6.15)$$

As already discussed in section 6.2.1, the gain K_2^* should assume a different value at each operating point to compensate for the nonlinear gain K_A in discontinuous mode. For sake of simplicity, a fixed value

$$K_2^* = 6.3 \qquad (6.16)$$

was experimentally determined and used in the present investigations. A critical performance occurs when the motor current changes from discontinuous to continuous. The resulting transient usually presents a current overshoot because a switch from the high controller gain, for discontinuous operation (Eq. 6.16), to the smaller gain, for continuous operation, Eq. (6.14), will not occur until the next sampling period, when the control algorithm will recognize the new current mode.

A current overshoot must be avoided, as high currents can melt the fast thyristor fuses. This difficulty was solved by decreasing K_2^* to 0.5 as soon as the current reference i_R assumes a value of a continuous current. This procedure makes the transient slower, but nearly without overshoot.

The algorithm is written in 8086-Assembler fix-point arithmetic. It occupies approximately 1K bytes of RAM and executes in 0.8 ms. The anti-wind-up algorithm suggested by Depping (1981) guarantees a good dynamic behaviour even when the firing angle limits are reached. The digital current controller also performs the necessary functions during a change in the direction of the load current, as already mentioned in section 6.2.1 for the analog controller. The required functions of the bridge selector (see Fig. 6.4) are easily programmed in assembly language, an advantage when compared with the hardware realization of the analog controller.

6.3.2 The Model Reference Adaptive Speed Controller

The model reference adaptive method has already been explained in chapter 5. The results that will be presented in chapter 7 are obtained with the following parameters:

reference model Eq. (5.16):

$$G_m(z^{-1}) = \frac{0.2}{1 - 0.8z^{-1}} z^{-1,} \qquad\qquad (6.17)$$

correction network Eq. (5.7):

$$G_c(z^{-1}) = \frac{0.07\,(1 - z^{-1})}{(1 - 0.5z^{-1})^2} z^{-1,} \qquad\qquad (6.18)$$

adaptation (Eqs. 5.36 to 5.39):

$$G = 100, \lambda_1 = 0.9, \lambda_2 = 0.7, \lambda_3 = 0.3, \rho = 1000, \gamma = 1000 . \qquad (6.19)$$

The reference model was chosen as a first order system with time constant of approximately 0.1 s and gain one. The discretization with a sampling time of 20 ms and using a zero order hold produces then Eq. (6.17).

The plant to be controlled with the model reference adaptive controller, i. e. the motor with inner current control loop, was identified at different operating points for a sampling time of 20 ms. A typical result, used for the synthesis, is given by

$$G_s(z^{-1}) = \frac{0.052 + 0.072z^{-1}}{1 - 0.96z^{-1} - 0.028z^{-2}} z^{-1},$$

This transfer function has a zero at -1.38 and therefore is nonminimum phase. A correction network is necessary. Choosing it as

$$G_c(z^{-1}) = \beta \frac{(1 - z^{-1})}{(1 - 0.5z^{-1})^2} z^{-1},$$

the augmented plant becomes minimum phase with the appropriate choice of the parameter β. The value $\beta = 0.07$ has produced good experimental results.

The parameters of the adaptation algorithm were initially obtained with digital simulation using the plant model described in Fig. 3.2. The results were confirmed with experimental tests.

It should be mentioned that adaptive control can not yet be applied routinely without an initial familiarization with the method. "A lot of design specifications including as much as possible a-priori-knowledge about the process must still be regarded " (Unbehauen, 1985).

The algorithm is written in PASCAL and occupies about 3K bytes. The assembler subprograms for floating point multiplication, division, subtraction, addition and comparison of data and parameter vectors described in chapter 4 make the use of the 8087 easy and the computation time of 13 ms nearly optimal. The sampling time is 20 ms.

6.4 The Completely Adaptive Hybrid Scheme

The completely adaptive hybrid scheme for model reference adaptive speed control is shown in Fig. 6.6 and for adaptive linear quadratic optimal or pole placement control in Fig. 6.7 . It consists of the analog dual-mode adaptive current controller described in section 6.2.1 and of an outer speed controller. The adaptive model reference speed controller is identical to that implemented in the digital scheme described in section 6.3.2. The adaptive linear quadratic optimal speed controller follows the principles proposed in section 5.4. Both controllers are fully integrated in the CADACS system discussed in chapter 4. They have analog inputs and an analog outputs and are well suited for replacement of different conventional control schemes. Further applications of these controllers with slightly changed operating interface are reported for control of a hydraulic positioning system (Unbehauen et al., 1989) and for regulation of electrical power and enthalpy of a 750 MW once-through boiler (Unbehauen et al., 1991).

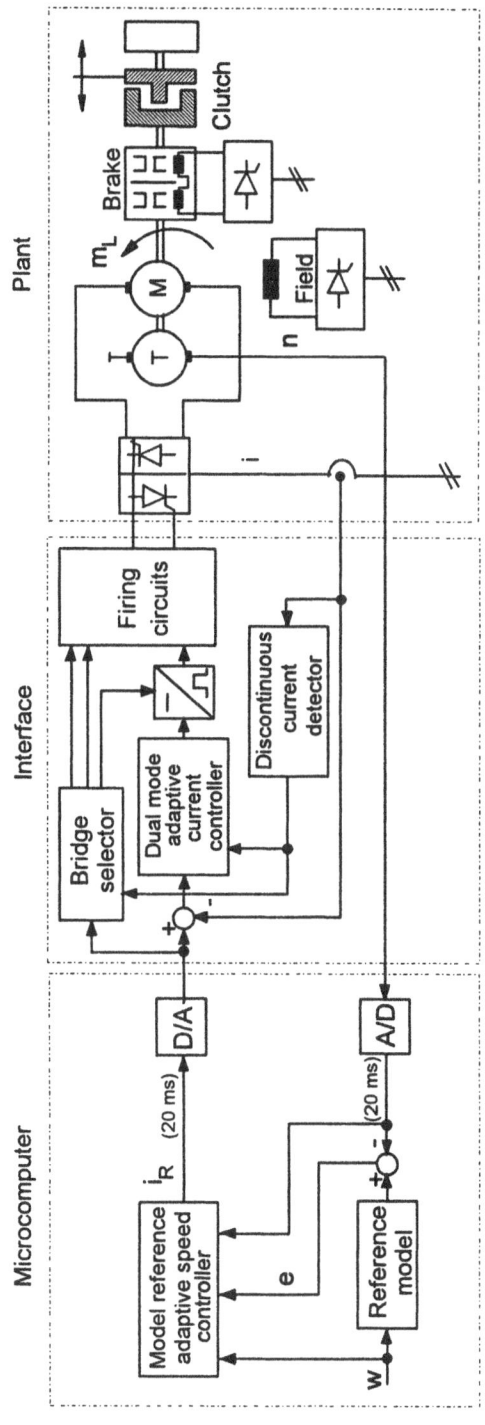

Fig. 6.6 The completely adaptive hybrid scheme with model reference speed controller (direct adaptive control principle)

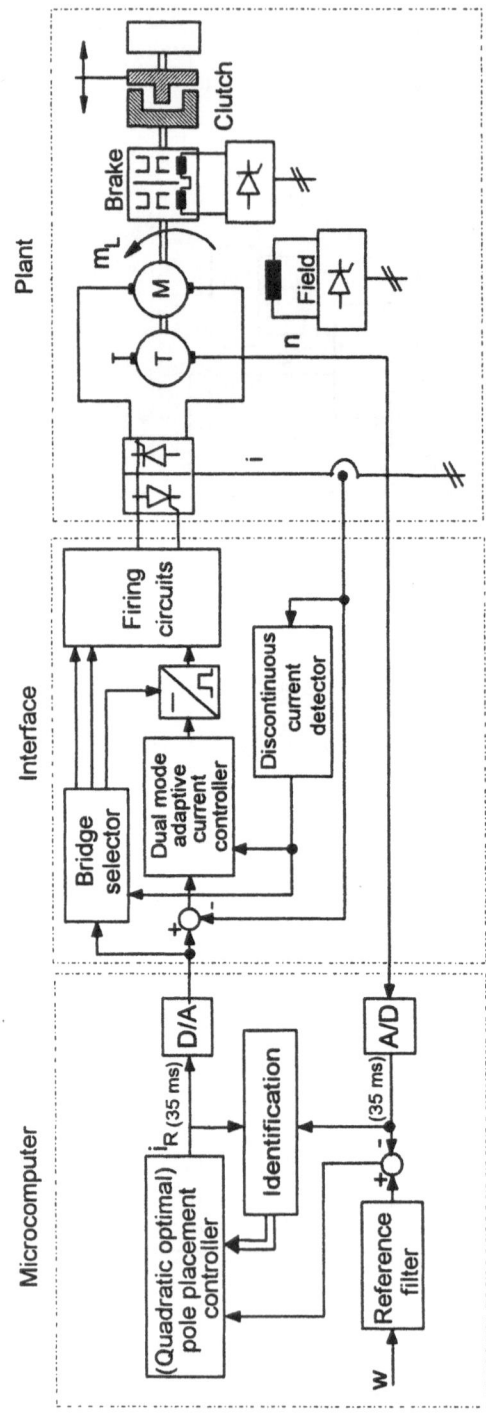

Fig. 6.7 The completely adaptive hybrid scheme with pole placement controller (indirect adaptive control principle)

6.5 Conclusion

The controllers used in the experimental tests were established in this chapter. The teachbox, described in chapter 4, serves for a simple start-up of the control circuit and allows to change the most important parameters. Alphanumeric and graphic display of the manipulated value, the speed variable and its control deviation can be used to monitor the actual state of the DC-motor.

A dialog program running on a PC connected by a serial communication line to the SBC allows alteration of all parameters, observation and documentation of important variables and parameters. Moreover, possibilities for digital simulation "in the loop" under real-time conditions permits a better familiarization with the adaptive algorithm and supports the developer in altering and improving the algorithms.

In the next chapter the performance of the various adaptive control schemes will finally be compared under different operating conditions.

6.6 References

Buxbaum, A. (1969), 'Regelung von Stromrichterantrieben bei lückendem und nichtlückendem Ankerstrom', *Techn. Mitt. AEG-Telefunken*, 59, pp. 348-352.

Buxbaum, A. and K. Schierau (1980), *'Berechnung von Regelkreisen der Antriebstechnik'*, AEG-Telefunken.

Bystron, K. (1979), *'Leistungselektronik, Vol II'*, Hanser Verlag, München.

Depping, F. (1981), 'Angepasste Sollwertvorgabe bei diskreten Reglern zur Vermeidung von Übersteuerungseffekten', *Regelungstechnik*, 29, pp.391-397.

Depping, F. and M. Voits. (1985), 'Kriterien der Betragsanschmiegung des Führungsfrequenzgangs als Leitlinie zur Optimierung von Abtastregelkreisen bei vereinfachtem Streckenmodell,' *Automatisierungstechnik*, 33, pp. 116-123 and 155-159.

Kessler, C. (1955), 'Über die Vorausberechnung optimal abgestimmter Regelkreise', *Regelungstechnik*, 3, pp. 40-48.

Kessler, C. (1958), 'Das symmetrische Optimum', *Regelungstechnik*, 6, pp. 395-400 and 432-436.

Lagasse, J. and R. Prajoux (1974), 'Behaviour of control systems including controlled converters, especially rectifiers: A review of existing theories', Proc. 1st IFAC Symp. Cont. in Power Electron. Elect. Drives, Düsseldorf, pp. 1-37.

Stephan, R.M., V. Hahn and H. Unbehauen 81988), 'Cascade adaptive speed control of a thyristor driven DC-motor', *Proc. IEE*, 135, pt. D, pp. 49-55.

Stephan, R.M., V. Hahn, J. Dastych and H. Unbehauen (1991), 'Adaptive and robust cascade schemes for thyristor driven DC-motor speed control', *Automatica*, 27, pp. 449-461.

Unbehauen, H. (1985), 'Theory and application of adaptive control', Prepr. 7th Conference on Digital Computer Applications to Process Control, Vienna, Austria, pp. 3-19.

Unbehauen, H., P. Du and U. Keuchel (1989), 'Application of microcomputer-based model reference adaptive control to a hydraulic positioning system', Proc. IFAC Symposium on Adaptive Control and Signal Processing, Glasgow, pp. 513-518.

Unbehauen, H., U. Keuchel and I. Kocaarslan (1991), 'Real-time adaptive control of electrical power and enthalpy for a 750 MW once-through boiler', Proc. IEE Control 91, Edinburgh, pp. 42-47.

Umland, J.W:, M. Safuddin (1990), 'Magnitude and symmetric optimum criterion for the design of linear control systems: What is it and how does it compare with the others ?', *IEEE Trans. on Ind. Appl.*, 26, pp. 489-497.

CHAPTER 7
EXPERIMENTAL RESULTS AND COMPARISONS

7.1 Introduction

The controllers defined in the last chapter were implemented experimentally using the hardware and software described in chapter 4. To evaluate their performance and to compare the results under practical operating conditions, laboratory tests have been carried out and will be discussed in the following sections.

7.2 Experimental Modelling

A direct adaptive control scheme is theoretically able to start without any information, the parameters could thus all be set to zero for start-up of the speed control loop in case of model reference adaptive control. In practice this would lead to undesirable values of the manipulated signal and could even produce damage in the controlled system. Thus at least an approximate model should be used for a priori parametrization of the adaptive control loop.

Without load, i.e., with an approximate mechanical time constant $T_H \approx 330$ms, for a sampling time of 50 ms one obtains, applying the off-line maximum likelihood parameter estimation method of the CADACS package to the measured input and output data, the discrete-time transfer function with zero order hold

$$G_1(z) = \frac{4.070_x 10^{-2} z^{-1} - 1.612_x 10^{-2} z^{-2}}{1 - 1.635 z^{-1} + 6.383_x 0^{-1} z^{-2}} \cdot \qquad (7.1)$$

Here the normalization 1000 rpm$\hat{=}$1 for the speed n and 10 A$\hat{=}$1 for the current $\bar{\text{i}}$ was used. $G_1(z)$ has poles and zeros according to Table 7.1.

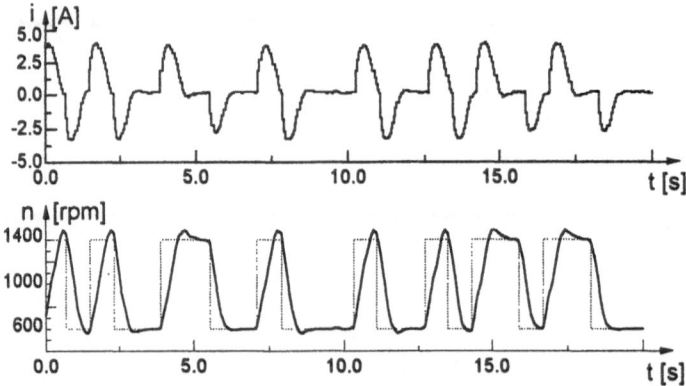

Fig. 7.1 Closed loop experiment for data acquisition (without load)

The rather high sampling time of 50 ms has been chosen for the modelling experiment to eliminate the noise, which influences the quality of the estimation results. For the pole placement controller with fixed parameters used throughout these experiments a sampling time of 3 ms could have been reached. But from experience and regarding the models deduced in chapter 3, this sampling rate can still be judged as quasi-continuous in respect of the dynamic behaviour of the speed control loop. Applying inverse z-transformation to the transfer function in Eq. (7.1), the associated continuous-time transfer function is given by

$$G_1(s) = \frac{6.835 \times 10^{-1}s + 1.220 \times 10^1}{s^2 + 8.976s + 1.585} \ . \tag{7.2}$$

Table 7.1 Poles and zeros of the model with motor time constant $T_H \approx 330$ms

discrete time		continuous time	
poles	zeros	poles	zeros
.644168	.396164	-8.79591	-17.8877
.991028		-.180248	

$G_1(s)$ can be used to derive discrete-time models for even smaller sampling times. Table 7.1 also shows the continuous time poles and zeros of the open loop system. Obviously, this model is minimum phase in discrete and continuous time description. But nevertheless provisions should be made in the model reference control algorithm in respect of the danger of discrete zeros with magnitude greater than one. There are mainly two reasons for this precaution:

- The bypass to the plant introduced by the correction network increases the robustness of the model adaptive control scheme as discussed in section 5.3. Thus, a lower adaptive gain can be chosen by specifying a covariance matrix with smaller norm in the estimation procedure. This reduces the variance of the parameters and keeps the control action and the closed loop dynamic behaviour smoother.

- The statistical dead-time of 5ms owing to the thyristor bridge will influence the estimation of the numerator parameters. If the controlled value is not sampled synchronously with the firing frequency of the AC/DC-converter, which is the case in our implementation, the numerator parameters will vary in time to approximate the changing dead time in terms of an allpass function.

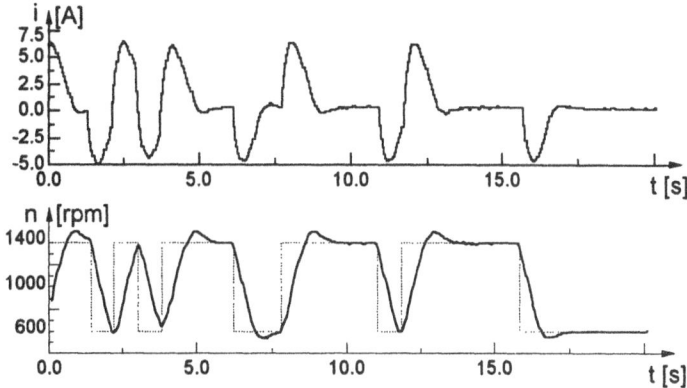

Fig. 7.2 Closed loop experiment for data acquisition (with load)

Table 7.2 Poles and zeros of the model with motor time constant $T_H \approx 700$ms

discrete time		continuous time	
poles	zeros	poles	zeros
.803151	.731660	-4.38425	-6.23387
.993083		-.138815	

With the additional load ($T_H \approx 700$ms) the transfer functions are given by

$$G_2(z) = \frac{2.306 \times 10^{-2} z^{-1} - 1.687 \times 10^{-2} z^{-2}}{1 - 1.796 z^{-1} + 7.976 \times 10^{-1} z^{-2}} \tag{7.3}$$

133

and

$$G_2(s) = \frac{4.436 \times 10^{-1}s + 2.765}{s^2 + 4.523s + 6.085 \times 10^{-1}} \cdot \qquad (7.4)$$

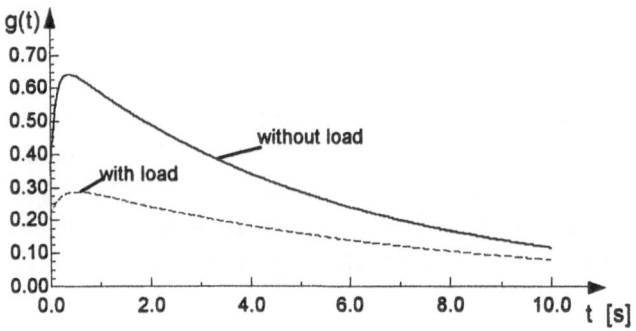

Fig. 7.3 Pulse responses for both experimental models

The model based on the experiment without load was used as initial parametrization for the experiments, regardless of the actual operating condition of the plant. The simulated pulse responses for both experimental models in Fig. 7.3 show the difference for these modes of operation. The pole near $z=1$ introduces nearly integral behaviour of the open loop system, the time constant of the associated first order lag is in the range 5..8 s. Thus the plant can be regarded as an approximate integrator for closed loop settling times less than 1 s.

7.3 Comparison of the Control Structures

The four cascade speed controllers compared here and already presented in chapter 6 are reproduced in Fig. 7.4. The entirely analog controller (Fig. 7.4a) is a standard commercially available unit, with dual-mode adaptive inner current loop. The other schemes were implemented and investigated during this work. The hardware of the entirely digital adaptive controller (Fig 7.4d) differs basically from the hybrid controllers assemblies (Fig. 7.4b,c). The latters have an analog current controller and the former a digital one. This difference has the following significant implications:

- The entirely digital solution necessitates the interface circuit between motor and microcomputer described in section 3.2.2. This circuitry represents an additional difficulty mainly due to the noisy motor/rectifier environment. On the other hand, the hybrid solution only needs a D/A converter to make the

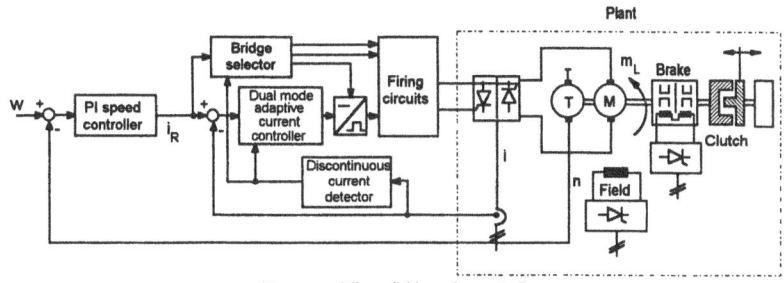

(a) commercially available analog controller

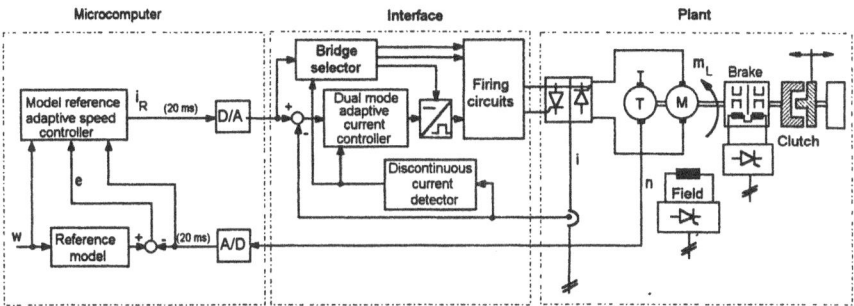

(b) completely adaptive hybrid scheme, model reference speed controller

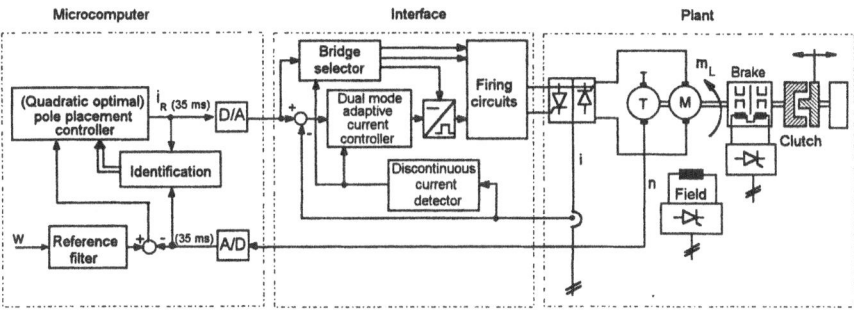

(c) completely adaptive hybrid scheme, pole placement speed controller

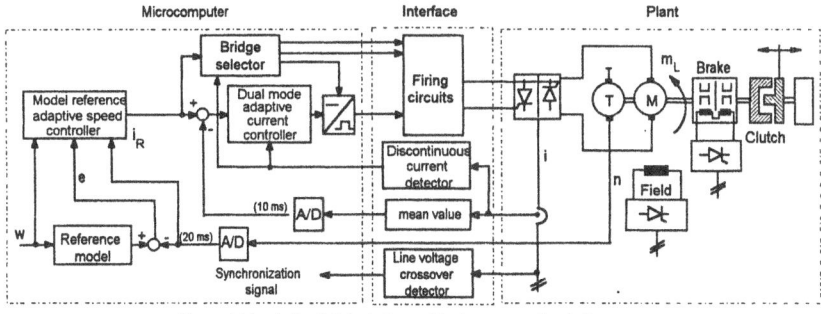

(d) completely adaptive digital scheme, model reference speed controller

Fig. 7.4 The cascade speed control schemes

digital current reference (i_R) available to the analog current controller, which is a standard commercial unit.

- The current loop, as already discussed, is highly non-linear. Moreover, the small armature time constant (30ms) requires high sampling rates for an optimal current control. These problems make the analog current controller more attractive than the digital one.

7.4 The Inner Current Loop

Two different current controllers are tested: the analog dual-mode adaptive and the digital dual-mode adaptive controllers. Their syntheses are respectively explained in sections 6.2.1 and 6.3.1. In Fig. 7.5 the dynamics of the current loop for step variations of the reference current value are shown for continuous and discontinuous current, as well as for the critical change from discontinuous to continuous current. The analog dual-mode adaptive controller gives in all conditions a rise time of approximately 50 ms and no overshoot. The digital controller, on the other hand, shows a greater variation of the rise time and a small current overshoot. This fact is expected according to the second observation made in section 7.3.

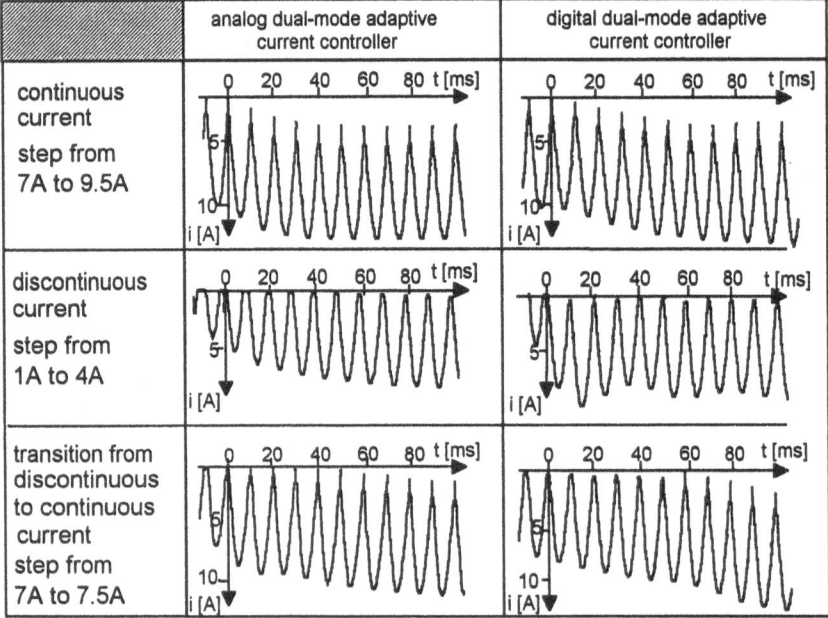

Fig. 7.5 Step responses for reference current variations under different conditions

The performance of the digital dual-mode adaptive controller is satisfactory when compared with the step response without adaptation shown in Fig. 7.6. The slow transient, with a rise time greater than 1.2s, can be obtained by switching off the adaptation of the analog or of the digital dual-mode adaptive controllers. This implies that the optimized PI-controller for continuous current actuates during the discontinuous current domain too.

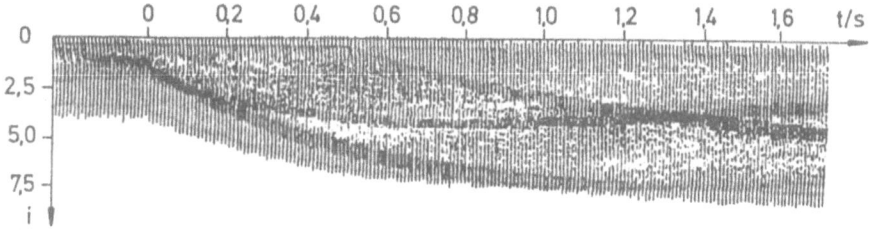

Fig. 7.6 Current step response in discontinuous current mode (1A → 4A); PI-current controller with parameters optimized for continuous current

The slow dynamics of the inner current loop can make the cascade speed control scheme unstable. In practice a limit cycle with a period of about 400ms and an amplitude greater than 200 rpm was observed, as shown in Fig. 7.7. This problem has been already discussed in literature, see for example Buxbaum and Schierau (1980).

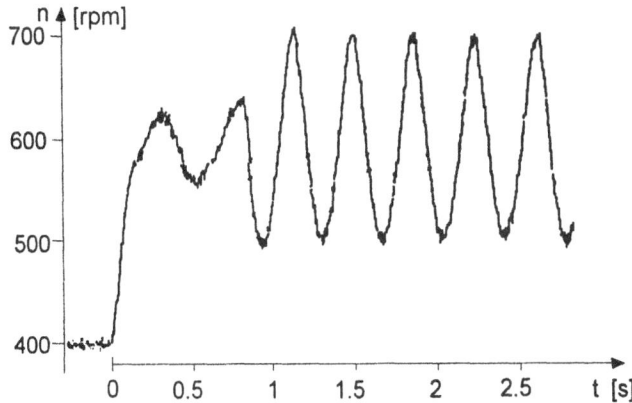

Fig. 7.7 Change of the reference speed from 1400 to 1600 rpm; cascade speed control scheme with PI-speed controller, no adaptation of the inner loop

The changes in the inner current loop in the case of discontinuous current can be also interpreted as a change in the parameters of the open loop system as seen from the speed controller. Thus adaptive speed controller should be sufficient in the case of discontinuous current without adaptive current controller. If an adaptive speed controller is used with appropriate weighting of the manipulated signal, the closed loop system will remain stable but the performance will decrease, as can be seen from Fig. 7.8. There is always a compromise between performance and sensitivity or robustness, when designing the controller. Actually, the weighting factor in the quadratic performance criterion of the linear optimal adaptive controller had to be increased by an order of magnitude to get the non oscillatory behaviour of the speed value. The manipulated variable still shows some oscillations due to the highly non-linear gain of the motor/rectifier system in discontinuous current mode.

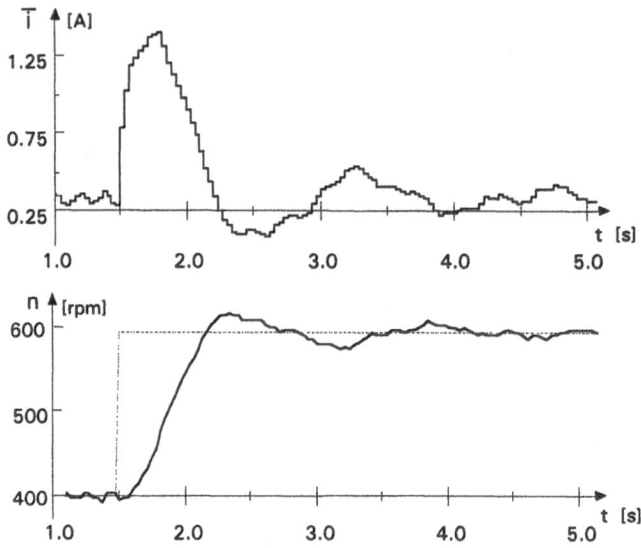

Fig. 7.8 Change of the reference speed from 1400 to 1600 rpm; adaptive speed control scheme with LQ speed controller ($\rho=1$), no adaptation of the inner loop

Regarding the results in Fig. 7.8 one can conclude that an adaptive speed controller is preferable in the case of a defective discontinuous current detector, because the speed control loop remains stable. But this should not be the only reason for the choice of an adaptive control strategy. The next section will examine the speed control loop in the case of a well adjusted and correctly working current controller.

7.5 The Cascade Speed Control Schemes

Four cascade speed control schemes are compared: (a) the commercially available analog scheme, (b) the completely adaptive model reference hybrid scheme, (c) the completely adaptive linear quadratic optimal hybrid scheme and (d) the completely adaptive model reference digital scheme. They are summarized in Fig. 7.4 and their design has been explained in chapter 6. The model of the plant, a thyristor-fed DC-motor, has already been presented in Figs. 3.2 and 3.3. The parameters and variables that appear in these figures are explained in chapter 3 and will often be mentioned during this section. If no comments are made in the text, the results obtained with the adaptive controllers are for well adapted parameters to set point variations.

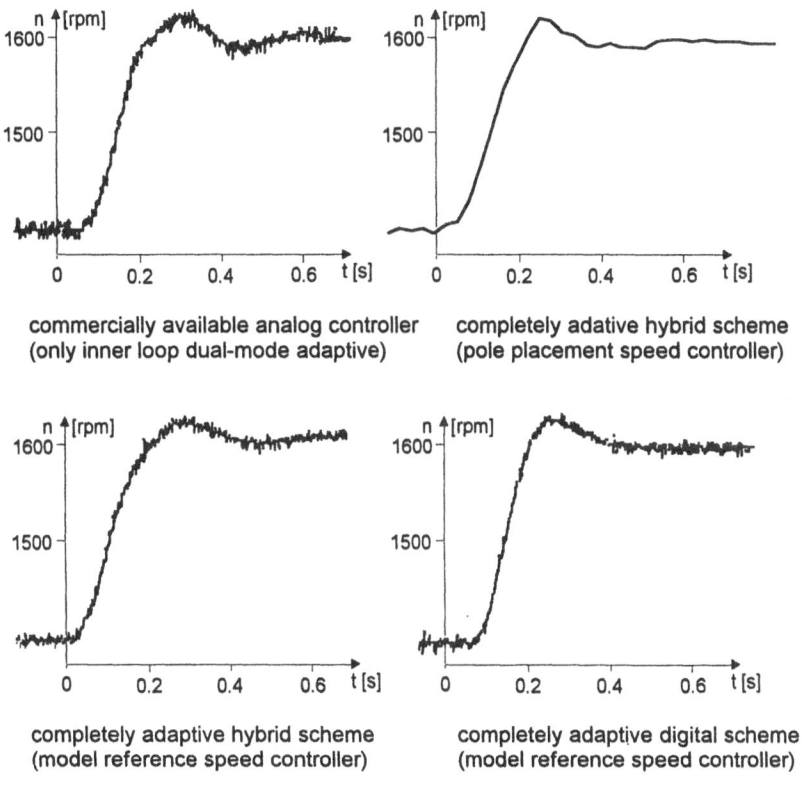

commercially available analog controller
(only inner loop dual-mode adaptive)

completely adative hybrid scheme
(pole placement speed controller)

completely adaptive hybrid scheme
(model reference speed controller)

completely adaptive digital scheme
(model reference speed controller)

Fig. 7.9 Step responses for a reference speed variation from 1400 to 1600 rpm.
Conditions: T_H=330 ms, ψ*=1.0, i_b=0.8A (discontinuous current) n $\overset{\Delta}{=}$ speed in rpm

Initially the speed step responses of the controllers are compared at an operating point given by nominal field excitation (ψ*=1), nominal moment of inertia (T_H=330 ms) and medium load (i_b=0.8A) of the eddy-current brake. Fig. 7.9 presents, for all

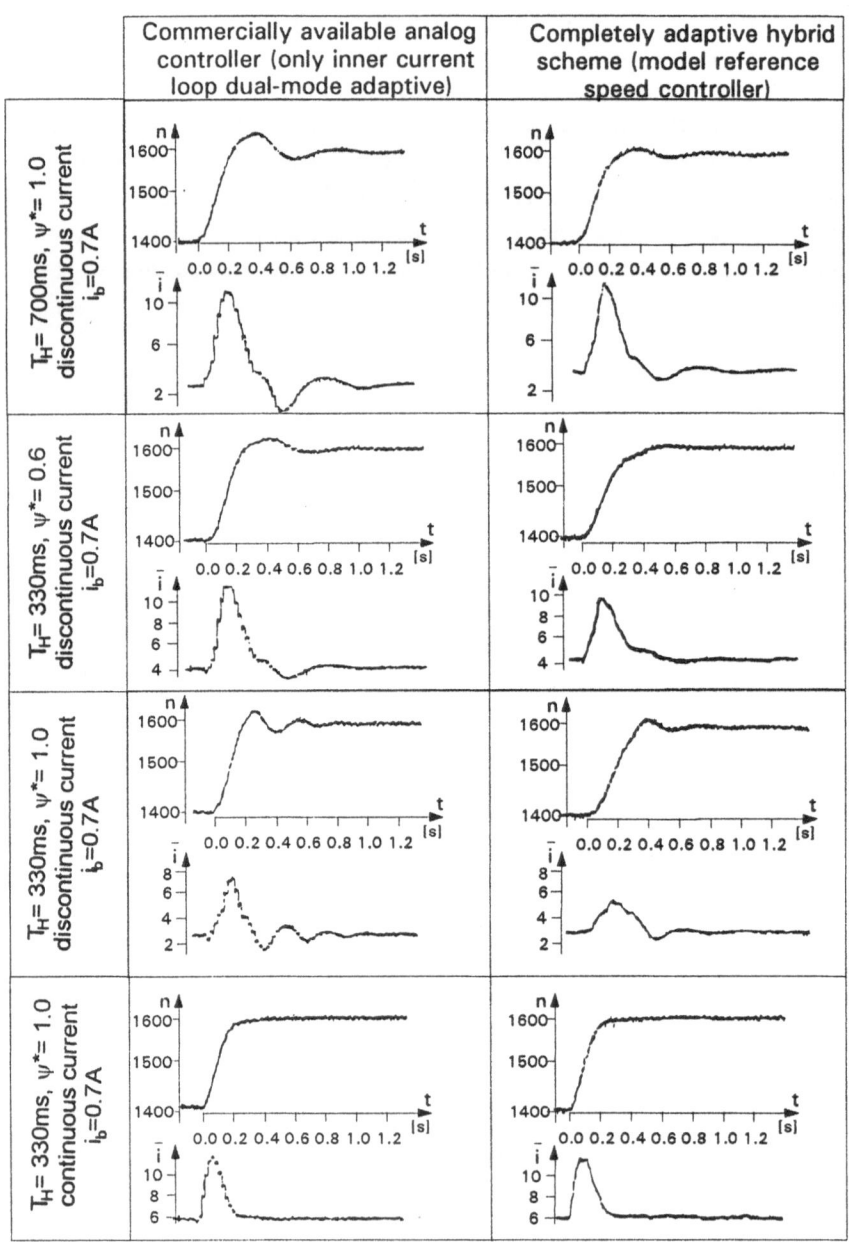

Fig. 7.10 Step responses for change of the reference speed from 1400 to 1600 rpm

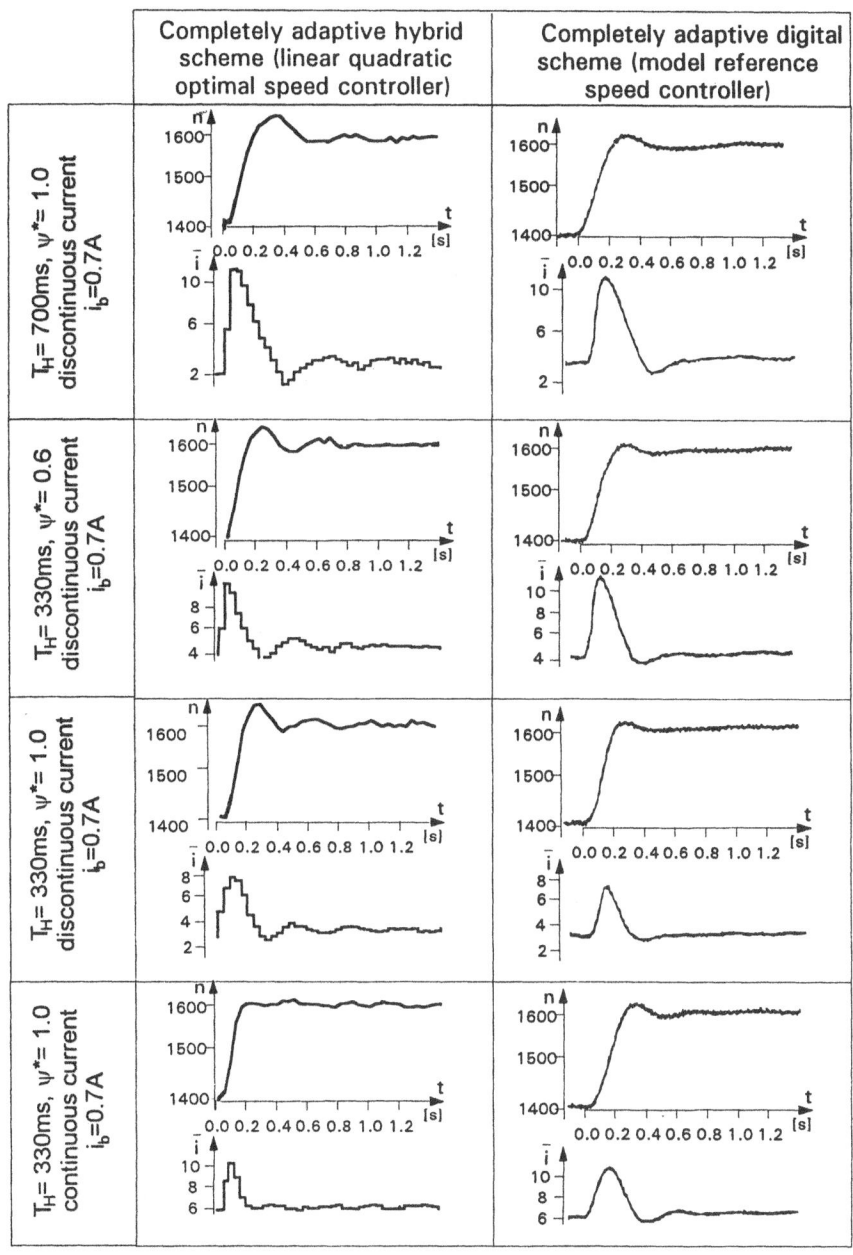

$n \stackrel{\Delta}{=}$ speed in rpm.

$\bar{i} \stackrel{\Delta}{=}$ armature current mean-value in A

schemes, an overshoot of approximately 10% and a settling time of 0.5 s. This shows the equivalence of the mentioned controllers at a well defined nominal operating point for tracking.

In Fig. 7.10 the controllers are compared at four different operating conditions. One can see that the completely adaptive model reference schemes provide a settling time of 0.5 s and an overshoot smaller than 10%. This is almost the performance given by the reference model (a first order system with time constant 0.1 s). The differences are due to the correction network, necessary in this application because the plant with inner current loop (i. e. the plant to be controlled with the model reference adaptive controller) has, at least temporarily, non-minimum phase characteristic considering its discrete time linear model with a sampling time of 20 ms (Stephan, 1985; Stephan, Hahn and Unbehauen, 1988). On the other hand, the commercially available controller and the adaptive linear quadratic optimal pole placement controller present at some operating conditions an overshoot greater than 10 % and a settling time greater than 0.5 s. The overshoots are mainly due to the integral action included in the control laws. The PI-speed controller generates closed loop poles which are conjugated complex and there are no degrees of freedom to attain an asymptotical dynamic behaviour of the speed control loop without degradation in the performance. In the case of the pole placement controller it is possible to position all poles on the real axis of the complex z-plane by switching off optimization and specifying real poles. But this will practically lead to reduced robustness and decrease the control performance as concerns disturbances. So it is advisable to choose a dominant conjugated complex pair of poles also in this case. The undesirable overshoot should be eliminated by introducing a prefilter and not by specifying real poles.

The effect of on abrupt change of the moment of inertia can be seen in Fig. 7.11. This perturbation serves also to demonstrate the effect of a pulse of load torque m_L, that occurs when the magnetic clutch is switched on and a mass, initially in repose, is set in motion. After this abrupt variation, the commercially available controller shows its characteristic response in the new operating condition. This response is similar to that of the adaptive linear quadratic optimal pole placement controller which also regulates the disturbance by the integral action of the control law. There is an unavoidable overshoot which increases the settling time. Experiments with feedforward of the estimated bias for the pole placement control scheme did not improve the performance. The model reference adaptive controllers perform well with negligible overshoot and small settling time. This type of mixed parametric change and disturbance can be handled in a nearly ideal way by the model reference adaptive controllers which do not incorporate the integral term. It is interesting to mention that changes of current direction, necessary during breaking conditions, are non-linear effects that render step responses from low to high speeds different from those from high to low speeds.

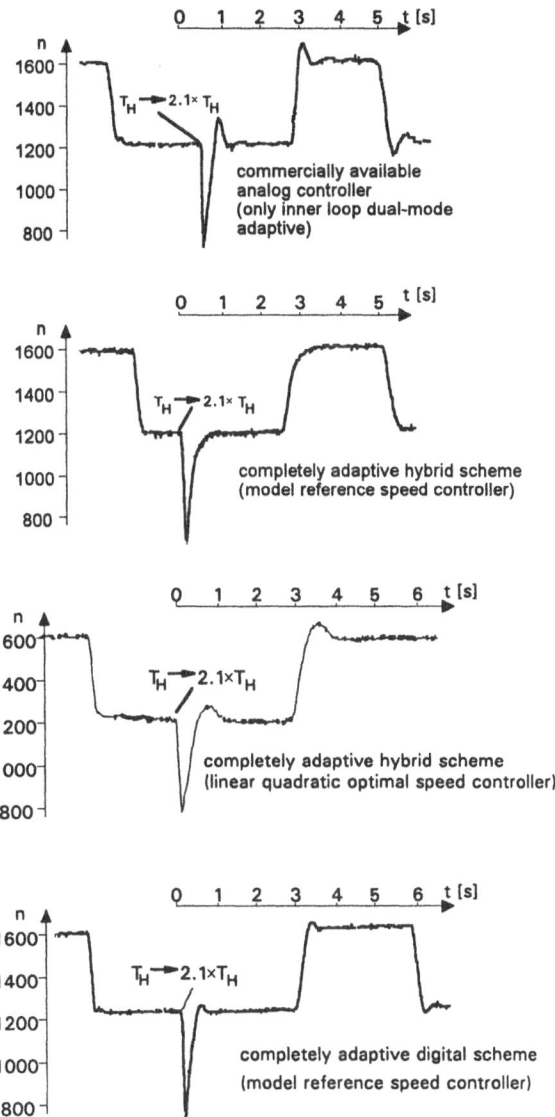

Fig. 7.11 Perturbation by means of a variation of the moment of inertia and subsequent changes of the reference speed.
Conditions: ψ^*=1.0, i_b=0.8A. n $\stackrel{\Delta}{=}$ speed in rpm

A variation of the load torque $\bar{m}_L$ can be obtained by altering the excitation current of the eddy-current brake i_b. Fig. 7.12 shows the speed performance in this case. The completely adaptive model reference controllers reject the load perturbation only due to the adaptation. Their response is slower than the response of the commercially available and of the linear quadratic optimal speed controllers, that have integral action.

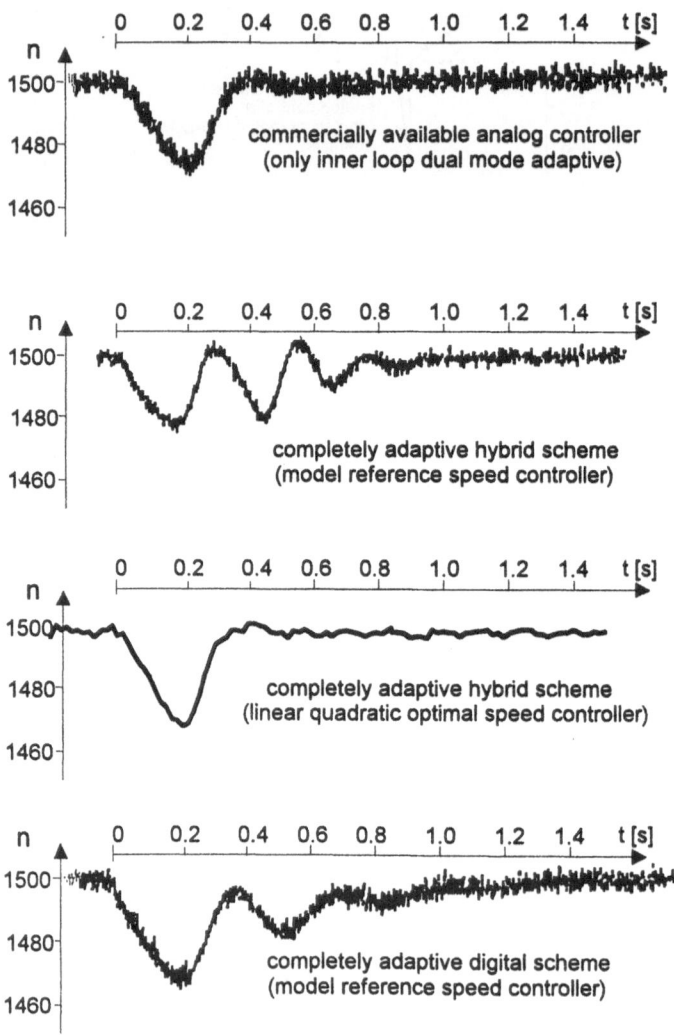

Fig. 7.12 Perturbation by means of a variation of the current of the eddy-current brake from i_b=0.7A to i_b=1.2A.. Conditions: T_H=330 ms, ψ*=1.0, n $\stackrel{\Delta}{=}$ speed in rpm

As was already described in section 5.2, an integral term could also be added to the model reference adaptive controllers, but experimental results have shown that such solution makes the reference current signal (the output of the model reference adaptive controllers) wobbly. This is expected in view of the approximate integral behaviour of the plant at the operating speed of 1500 rpm. As can be seen from Fig. 3.3, for n*=0.79, the coefficient B is nearly zero and therefore, according to Fig. 3.2, the electro-mechanical part of the plant behaves as an integrator. The speed control loop then, with two integrators, presents a structural instability to the adaptive controller, thus justifying explains the wobbly characteristic mentioned above. This problem can not be solved specifying closed loop poles in the design of the correction network because the undesired behaviour is often introduced by zeros in the disturbance transfer function of the closed loop system. A possible solution is the factorization of the plant polynomial prior to the on-line design of the correction network regarding desired and undesired regions for poles and zeros in the complex z-plane as proposed by Wiemer et al. (1988). But this factorization is not of the spectral type so that there is not generally available a fast procedure for on-line calculation of this factorization.

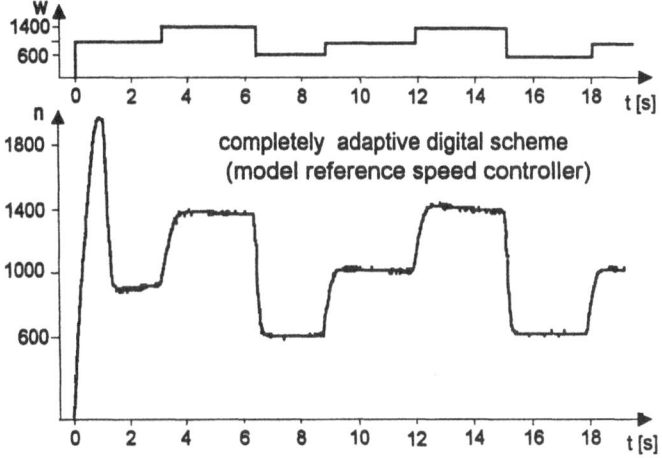

Fig. 7.13 Model reference controller during adaptation of the controller parameters. Conditions: T_H=330 ms, ψ*=1.0, i_b=0.8A; w $\triangleq$ reference speed, n $\triangleq$ speed in rpm

Moreover, a reference model for set point variations, and not for load perturbations, is given for the model reference adaptive scheme. This fundamental design consideration handicaps the step response for perturbations and is also responsible for slower parameters adaptation in the case of load perturbations than in the case of set point variations

The parameter adaptation for set point variations can be judged by the results shown in Fig. 7.13. All parameters of the model reference adaptive controller are initially set to zero and the adaptive algorithm is then started with a reference speed of 1000 rpm. Then the reference speed is varied between 1400 rpm, 600 rpm and 1000 rpm, furnishing an identification signal known as ternary (Unbehauen and Ganti, 1987). Satisfactory responses are already obtained during the second step. Thus, the controller parameters are adapted.

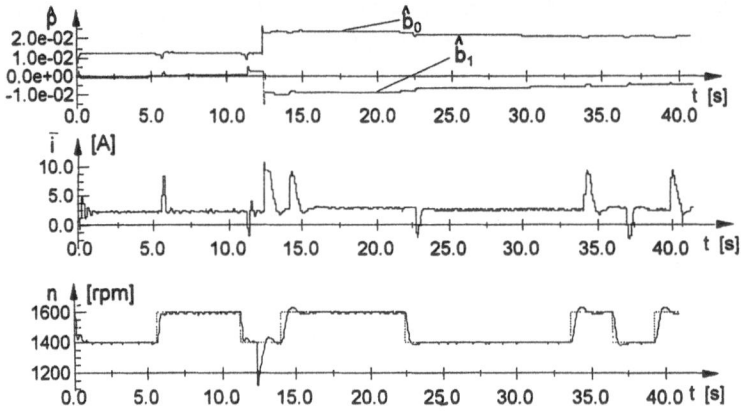

Fig. 7.14 Controlled variable, manipulated variable and estimated parameters during adaptation of the controller parameters, conditions as in Fig. 7.13

The adaptation of parameters during the operation of the indirect adaptive pole placement are illustrated in Fig. 7.14 . Here, the parameters of the discrete transfer function of the plant are estimated. The start-up was made with detuned parameters (setting the parameters to zero is not allowed as in direct adaptive control). Also in this case a good nominal behaviour is already reached in the second step change of the pseudo-random-binary reference signal. After 12 s the clutch is switched on so that the moment of inertia changes by a factor of 2.1, introducing a step change in the parameters. As can be deduced from the course of the manipulated signal, the dynamic behaviour of the system is directionally dependent because a change of the current direction occurs. This non-linear effect leads to a continuous variation of the parameters.

146

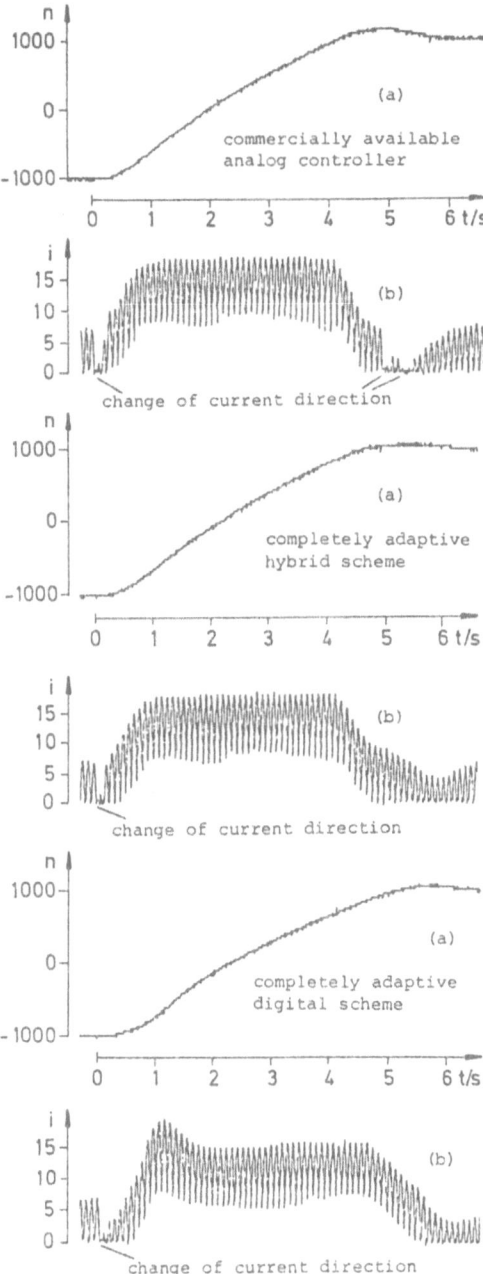

Fig. 7.15 Change of speed direction (four-quadrant operation). Conditions: T_H=330 ms, ψ^*=1.0, i_b=0.8A. (a) speed in rpm; (b) current absolute value in A

Finally, Fig. 7.15 shows the change of speed direction. This dynamic behaviour is accompanied with changes of the current direction. The functions necessary to accomplish these changes are discussed in section 6.2.1 (bridge selection) and are implemented by software in the digital adaptive scheme, and by standard hardware, for the hybrid adaptive and commercially available controllers. The first type shows a current overshoot, whereas the two last mentioned do not. This confirms, once again, the second observation made in section 7.4. Moreover, during these transients, the limits of the reference current are reached and a wind-up reset is necessary in the digital implementation.

The spikes that can be observed in some of the curves, especially in Fig. 7.12, are due to the electrical noise present in the motor-rectifier-tachogenerator environment. The experiments involving the pole-placement controller do not bring forth this high frequency noise due to the greater sampling time and the use of a special filter in the interface unit.

7.6 Conclusion

The results obtained in this chapter show that the model reference adaptive control concept can be applied with success for the speed control of a DC-motor. For step responses or disturbances, that can be interpreted as parameter variation, such as in the case of changes in the moment of inertia, this control scheme is superior to the other control algorithms with fixed or adaptive control law.

Despite higher overshoot owing to the integral action included in the control law, the linear quadratic optimal adaptive pole placement controller performs nearly as well as the model reference adaptive controller for tracking and parametric changes. For regulation of disturbances injected by the eddy-current brake the closed loop behaviour is as good as with the PI controller.

The estimation procedures with time-varying weighting factor performed well and robust. Neither estimator wind-up or covariance blowup nor a degradation in parameter tracking could be observed even in long-term experiments of several hours, with changes in operating conditions and varying load.

The combination of the inner dual-mode adaptive current controller based on parameter scheduling and an adaptive speed controller with controlled perpetual adaptation proved to be an optimal control structure for thyristor fed DC-motors. The inner current loop copes with all the effects introduced by the mode of current flow and contributes to the regulation of disturbances, so that during the design of the outer loop one can concentrate mainly on the tracking behaviour, thus obtaining an overall good control performance for all operating conditions.

7.7 References

Buxbaum, A. and K. Schierau (1980), *'Berechnung von Regelkreisen der Antriebstechnik'*, AEG-Telefunken.

Stephan, R.M. (1985), 'Adaptive abd robust cascade schemes for thyristor driven DC-motor speed control using a microcomputer', Dr.-Ing. Thesis, Ruhr-University Bochum.

Stephan, R.M., V. Hahn and H. Unbehauen (1988), 'Cascade adaptive speed control of a thyristor driven DC-motor', *IEE Proc.*, 135, pp. 49-55.

Unbehauen, H. and G.P. Rao (1987), *'Continuous-Time System Identification'*, North-Holland, Amsterdam.

Wiemer, P., G. Olejua Torres and H. Unbehauen (1988), 'A robust adaptive controller for systems with arbitrary zeros', Prepr. Int. IEE Conf. CONTROL 88, Oxford, pp. 598-603.

CHAPTER 8
SURVEY AND CONCLUSION

The material presented in the last seven chapters contributes to the experimental realization and to the critical judgement of new adaptive control concepts recently proposed. The plant to be controlled consists of a separately excited DC-motor fed by a single-phase fully-controlled dual-converter. Variations of the moment of inertia, field excitation and load torque are possible. Moreover, the single-phase supply causes the discontinuous current domain to be wide. It is known that when the current is continuous, the motor armature can be regarded as a first order system, but when the current is discontinuous, it changes into a nearly nonlinear gain system. All these variations make the application of adaptive solutions for the motor speed control attractive. On the other hand, the fast electro-mechanical system and the complex adaptive algorithm challenge the use of new developments of microelectronics.

As is common practice in control engineering the investigation was performed in the following sequence: modelling, identification, simulation, design and implementation. The mean-value model of a thyristor-fed DC-motor is already heuristically known. Here a mathematical explanation of this mean-value model was proposed, enlightening important aspects. The identification was made with well established methods. A cascade strategy was chosen for the DC-motor speed control design, and four schemes were compared:

- A commercially available analog scheme: An analog dual-mode adaptive inner current loop is cascaded with a simple analog PI-speed control loop.

- A completely adaptive digital scheme: A digital dual-mode adaptive inner current loop is cascaded with a model reference adaptive speed control loop.

- A completely adaptive hybrid scheme. The commercially available analog inner current loop is cascaded with a digital model reference adaptive outer loop.

- A completely adaptive hybrid scheme. The commercially available analog inner current loop is cascaded with a digital linear quadratic optimal pole placement speed controller.

The digital controllers had been implemented on a 16-bit single board microcomputer and most of the software was written in Pascal. An arithmetic coprocessor added the possibility of fast floating-point computation, limiting the need of scaled algorithms. But, for time critical tasks, assembler subroutines and fix-point arithmetic were used. Moreover, the software development was not restricted to the present application. A monitor, numerous subroutines and some programs provide the dialog between the mentioned single board computer and a personal computer, used as development system, comfortable and general. In fact, nowadays other digital controllers are being developed using these facilities.

The obtained results showed that the adaptive control concept can be applied with success for the speed control of a DC-motor.

Finally, this work intended to be a bridge between control theory and practice. This bridge had two great pillars: power electronics and micro electronics. Therefore, the abstract and the real, the big and the small came here face to face.

Subject Index

A

A/D-D/A converter, 38
AC-DC converter, 13
adaptation law, 73, 74
adaptive control, 67
adaptive control law, 72
analog speed controller, 118
arithmetic coprocessor, 59
armature current, 13, 18
armature gain, 21
armature model, 17, 21
armature power rectifier, 28
armature time constant, 14
armature voltage, 13, 18, 29
augmented error signal, 73
augmented plant, 78

B

back emf, 16
back voltage, 13, 15, 18, 120
benchmark, 37
Betrags-Optimum, 22, 120
bias estimation, 107
bridge selector, 121

C

CADACS, 47
calculation of the controller, 111
cascade speed control, 139
cascaded speed control, 1
certainty equivalence principle, 97, 102
characteristic polynomial, 97
constant trace, 83
continuous current mode, 19, 22, 120
control deviation, 77
control engineering software, 63
control error, 91

control law, 76, 82, 87
control software, 63
control structures, 134
controller synthesis, 76
correction network, 69, 72, 77, 79
covariance matrix, 83, 94
CPU-board, 37
current conduction modes, 21
current control, 21
current control loop, 23
current controller, 24
current loop, 120
current mean-value, 19
current reference limiter, 122
current transducer, 26

D

D/A-A/D-converter, 43
data base, 48
data vector, 93, 94, 106
DC-machine, 28
DC-motor, 26
DC-motor model, 13
dead times, 20, 98
decreasing gain estimation, 74
design equations, 76, 78, 92
deterministic disturbance, 81, 100, 103
device driver, 49
dialog program, 33
digital adaptive control, 134
digital current control, 122
digital speed control, 1
Diophantine equation, 90, 98, 104
direct adaptive control, 68
direct speed control, 1
discontinuous current mode, 13, 18, 22, 23, 120
disturbance feedforward, 102
disturbance filter, 100

dual-mode adaptive, 22, 24, 136

E

EC-bus, 36
ECB, 36
eddy-current brake, 28, 29
error signal, 71

F

FELDBUS, 65
field current, 29
field excitation, 26, 28, 121
filtered error, 81, 82
firing angle, 13, 19, 29
firing pulse, 121
forgetting factor, 84
friction coefficient, 16

G

gradient vector, 94, 97

H

hybrid adaptive control, 134

I

I-controller, 22
I/O library, 42
indirect adaptive control, 68, 85, 97
initial parametrization, 131
integral action, 70
integral control law, 106
internal model principle, 104
interrupt controller, 34
interval timer, 34
inverse cosine control, 29

K

Kalman gain vector, 83

L

limited manipulated signal, 108
linear model following control, 5
linear quadratic optimal control, 75
linear quadratic optimal control law,
118
linear quadratic regulator, 91
load torque, 16, 26, 29

M

machine constant, 16
magnetic clutch, 28
magnetic flux, 16
mailbox I/O, 50
maximum likelihood parameter
estimation, 131
mean-value model, 20, 24
memory map, 42, 43
microcomputer, 34
model behaviour, 76
model error, 68, 81
model reference adaptive control, 5,
68, 75, 82, 118, 125
multi-I/O-board, 38
MULTIBUS, 42

N

non-minimum systems, 69
numeric coprocessor, 43

O

observer poles, 113
observer polynomial, 87, 90
open-loop transfer function, 22

P

parallel interface, 34
parameter identification, 69
parameter update, 73
parameter vector, 73, 82, 93, 94, 106
performance criterion, 91

perpetual adaptation, 67
PI-controller, 22
PI-speed regulator, 118
pole placement control, 85
power-on reset, 44
prediction error, 83, 94, 107
PROFIBUS, 65
program development, 56

R

RAM board, 38
real-time monitor, 53
real-time program, 33, 53
real-time scheduler, 45, 46
real-time system, 34
real-time task, 47
real-time toolbox, 57
rectifier dead time, 22
recursive least squares, 83
recursive maximum likelihood
method, 93
recursive prediction error method, 93
reduced order observer, 89
reference model, 71, 76
RLS-algorithm, 83
robustness, 75
ROM-BIOS, 41

S

separation principle, 90
serial communication, 34
signal vector, 73, 82
single board computer, 34, 41
single-phase bridge, 13
spectral factorization, 75, 80, 92, 96,
97, 104, 105
speed, 20
speed controller, 28
speed transducer, 26
square-root filter algorithm, 83
stability, 74
stability theory, 75
state feedback, 5
state observer, 87
state space equation, 85

state vector, 85
steady-state behaviour, 77
step disturbances, 72
stochastic approximation, 73
stochastic disturbance, 92, 94, 99
supervisory system, 68
supply voltage, 15
support software, 55
Symmetrisches Optimum, 121
synthesis of a pole placement control,
91

T

tachogenerator, 28
teachbox, 36, 40, 43
technical data, 31
thyristor firing, 29
thyristor power supply, 14
torque, 16
torque balance, 16

U

user interface, 54

V

vector instruction set, 59

W

weighted RLS methods, 95
weighting factor, 83, 84